Foundations of Mathematics

Volume 1
Sets and Number Systems

Angela G. Shirley

Copyright © 2017 by Angela G. Shirley

All rights reserved. No part of this publication may be reproduced, stored in a retrieval system, or transmitted in any form or by any means—electronic, mechanical, photocopy, recording, or any other—without prior permission of the copyright holder. The only exception is brief quotations in printed reviews.

Foundations of Mathematics
Volume 1: Sets and Number Systems
By Angela G. Shirley

ISBN-10: 1537028618
ISBN-13: 978-1537028613

BISAC: Mathematics/Algebra/Abstract

AS Publishing, St. Augustine, Trinidad, W.I.
Email: aspublishing.tt@gmail.com

Front Cover Image: *The School of Athens* by Raphael, 1509-1511
—a celebration of ancient Greek philosophy as the foundation of Renaissance learning.

Printed by CreateSpace

TO MY STUDENTS

*"We are what we repeatedly do.
Excellence then, is not an act, but a habit."*

—Aristotle

CONTENTS

FOREWORD ix
PREFACE... xi
ACKNOWLEDGEMENTS xiii

1. ELEMENTARY LOGIC 1
    1.1 STATEMENTS 1
    1.2. CONNECTIVES 2
    1.3 CONVERSE AND CONTRAPOSITIVE.. 7
    1.4 NECESSARY AND SUFFICIENT CONDITIONS 9
    1.5 TAUTOLOGIES AND CONTRADICTIONS 11
    1.6 QUANTIFIERS 12
    1.7 VALID ARGUMENTS 13
    1.8 PROOF BY CONTRADICTION 20
    1.9 FEATURE MATHEMATICIAN 23
    1.10 CHAPTER EXERCISES 24

2. SETS 27
    2.1 BASIC IDEAS AND NOTATIONS 27
    2.2 OPERATIONS ON SETS 30
    2.3 FORMAL PROOF OF SET EQUALITIES 32
    2.4 SET ALGEBRA 37
    2.5 APPLICATION OF SET THEORY TO ARGUMENTS 40
    2.6 FEATURE MATHEMATICIAN 45
    2.7 CHAPTER EXERCISES... 46

3. BINARY OPERATIONS AND EQUIVALENCE RELATIONS 49

3.1 BINARY OPERATIONS 49
3.2 SOME PROPERTIES OF REAL NUMBERS 54
3.3 WORKED EXAMPLES–BINARY OPERATIONS 55
3.4 EQUIVALENCE RELATIONS 61
3.5 EQUIVALENCE CLASSES 66
3.6 WORKED EXAMPLES–EQUIVALENCE RELATIONS 69
3.7 FEATURE MATHEMATICIAN 71
3.8 CHAPTER EXERCISES 72

4. FUNCTIONS 75

4.1 DEFINITION OF A FUNCTION 75
4.2 ONTO FUNCTIONS 78
4.3 ONE TO ONE FUNCTIONS 79
4.3 ONE TO ONE CORRESPONDENCE 81
4.4 COMPOSITION OF FUNCTIONS 84
4.5 INVERSE FUNCTIONS 85
4.6 FEATURE MATHEMATICIAN 89
4.7 CHAPTER EXERCISES 90

5. NATURAL NUMBERS 91

5.1 THE NUMBER SYSTEMS 91
5.2 INDUCTIVE SETS 93
5.3 PRINCIPLE OF MATHEMATICAL INDUCTION 93
5.4 FEATURE MATHEMATICIAN 99
5.5 CHAPTER EXERCISES 100

6. REAL NUMBERS 101
6.1 THE 'LESS THAN' RELATION 101
6.2 ABSOLUTE VALUE 102
6.3 PROVING REAL NUMBER INEQUALITIES 106
6.4 SOLVING REAL NUMBER INEQUALITIES 109
6.5 INEQUALITIES INVOLVING ABSOLUTE VALUE 111
6.6 FEATURE MATHEMATICIAN 113
6.7 CHAPTER EXERCISES 114

7. INTEGERS, RATIONALS AND IRRATIONALS 117
7.1 EXTENSION OF N TO Z 117
7.2 SOME PROPERTIES OF INTEGERS 119
7.3 PRIME NUMBERS 121
7.4 FUNDAMENTAL THEOREM OF ARITHMETIC 122
7.5 LINEAR DIOPHANTINE EQUATIONS 123
7.6 RATIONAL NUMBERS 125
7.7 IRRATIONAL NUMBERS 127
7.8 FEATURE MATHEMATICIAN 129
7.9 CHAPTER EXERCISES 130

APPENDIX 1— COURSE DESCRIPTION 133

APPENDIX 2— PAST EXAMINATION PAPERS 137

APPENDIX 3— SOLUTIONS TO PAST EXAM PAPERS 149

FUTURE MATHEMATICIAN 185

ABOUT THE AUTHOR 187

FOREWORD

It is highly desirable that students of Mathematics, at least at the university level, be exposed to the fundamental principles on which the subject is based. It is therefore absolutely necessary for them to be acquainted with, and hopefully understand, some of the vital topics in the most fundamental area of Mathematics—mathematical logic. Only with such a knowledge, would it be possible for them to appreciate and understand the development and usage of the most basic element in Mathematics—the set.

This book, *Foundations of Mathematics: Volume 1, Sets and Number Systems*, gives a presentation of topics in Mathematics, in the desired logical sequence. Firstly, it deals with the minimum essential tools of Mathematical Logic, which are used for arriving at mathematically acceptable conclusions. The basic definitions and ideas in Set Theory, and their logical extensions to vital topics (some of which students are superficially familiar with), are then presented in a logical manner.

The text introduces the various number systems and their interesting properties. The unique properties of each number system are highlighted. The importance of these properties is then reinforced by means of a variety of worked examples, discussions and exercises.

The book is very student friendly. Each chapter contains useful exercises and relevant references. Throughout the book there are candid comments and questions which facilitate a proper understanding of the material presented. Another special feature of this book is the historical information given about the great founding fathers of Mathematics, via its "Feature Mathematician", at the end of each chapter.

Foundations of Mathematics: Volume 1, Sets and Number Systems covers all the topics in the syllabus (Appendix 1) for the M1152 course, given at The University of the West

Indies, St. Augustine. It also provides detailed solutions to some of the past examinations papers (Appendices 2 and 3). It is certainly a welcomed addition to the available literature.

Edward J. Farrell

Emeritus Professor of Mathematics

Department of Mathematics and Statistics

The University of the West Indies

St. Augustine, TRINIDAD

PREFACE

One of the greatest mathematicians of all time is reported to have said, "Mathematics reveals its secrets only to those who approach it with pure love, for its own beauty." *Foundations of Mathematics* is devoted to awakening a love for Mathematics among first year university students and is a product of the teaching of Introductory Mathematics at the University of the West Indies (UWI), St Augustine, for over twenty years. It is a two-volume series—Volume 1, *Sets and Number Systems,* introduces the student to concepts in Abstract Algebra, while Volume 2 focuses on Linear Algebra. *Sets and Number Systems* deals with basic ideas in logic, sets, relations, functions, binary operations, and the number systems. It is intended to cover a three credit, one semester course in Sets and Number Systems (coded as MATH 1152 at UWI).

My objective is to get the student to be able to think and express him/herself mathematically, to argue logically, and to appreciate the basic concepts on which Mathematics is built. The emphasis is on understanding these concepts and learning the language of mathematics rather than on mechanical computation. *Sets and Number Systems* lays a firm foundation for second year courses in Mathematics and Mathematics-related Sciences. It is written not only for students on the UWI, St. Augustine Campus, where I teach, but also for students on the Cave Hill and Mona campuses, whose courses are harmonized with ours. Indeed, the text can help any student pursuing a similar course in Mathematics and will serve as a valuable resource for graduate students who help with the teaching of the course.

The material presented in *Sets and Number Systems* is a collection of notes prepared over my time of teaching the course. It is simply, but thoroughly, written; and contains many worked-examples and exercises to give the student ample practice and aid in understanding how solutions should be written. Sources used in compiling the notes are given in the references at the end of each chapter and can serve as additional reading. The chapters may be studied in the order presented, as this is found to be most helpful for teaching purposes, but some may prefer to do Chapter 4 before Chapter 3 and Chapter 7 before Chapter 6 as an equally logical sequence. Each chapter also features a famous mathematician whose work in Mathematics was foundational. The life stories of these heroes of Mathematics can inspire us to do great things.

The appendix contains material specific to the MATH 1152 course offered at St Augustine. Appendix 1 gives the course description and Appendix 2 gives some of the department's past examination papers. These are also posted on the University's website. Appendix 3 gives solutions to these papers. The student must note, however, that Mathematics is not a spectator sport. You cannot learn Mathematics merely by *looking*. Rather, you must *do* the exercises yourself. You must earnestly try the exam questions before looking at the solutions. Otherwise, you might think that you can solve the problems and find it to be a different story when you actually start writing your answer. If you have difficulty in completing an exercise, you should go back to that topic and study the material again. Answers are intentionally not provided for the Chapter Exercises. You can practice these on your own and ask about them in your tutorials. A final page features the math student as a future mathematician.

This publication—*Sets and Number Systems*— is in no way intended to replace classes but only to reinforce the teaching and to lessen the burden of note-taking. By making all these materials available to the student it is hoped that you will have a greater appreciation and love for Mathematics and that we will see a dramatic increase in the pass rate for the course. May no one who enters the pages of this book, leave ignorant of Algebra!

Angela G. Shirley

St. Augustine, Trinidad
August, 2016

ACKNOWLEDGEMENTS

I would like to thank the University of the West Indies (UWI), St Augustine, and the Department of Mathematics & Statistics (DMS) for their support of this work and for granting me the time to complete it. Thanks to Edward J. Farrell, Emeritus Professor of Mathematics, UWI, for his meticulous editing of the material and for writing the Foreword. Professor Farrell was a primary developer of the department's introductory mathematics courses and displayed keen enthusiasm in seeing this project completed. Thanks to Dr. Shanaz Wahid, Senior Lecturer, DMS, for his gracious help in reviewing the work and for his hearty endorsement.

A special thanks to graduate student, Khatiza Mohammed, for her assistance in typing the manuscript, and for the beautiful Visio drawings. Though she is not a mathematics student, her command of Equation Editor is amazing. Thanks to DMS graduate students, Anna-Keren McMayo and Kelli Chang, for their photo of Future Mathematician. Thanks, also, to all my first-year students for providing years of opportunity in refining these notes and for the appreciation they have always shown for them.

I thank my family for their continuous encouragement.

Above all, thanks to the Author and Finisher of all good things.

Angela G. Shirley

CHAPTER ONE

ELEMENTARY LOGIC

In this chapter we look at some basic ideas of Mathematical Logic, including: Propositional Statements, the Connectives *and, or, not, if . . . then,* and *if and only if,* Truth Tables, Logical Equivalence, Converse and Contrapositive of an Implication, Necessary and Sufficient Conditions, Tautology and Contradiction, Quantifiers, Validity of Arguments and Proof by Contradiction. The aim is to teach you to reason logically.

1.1 STATEMENTS

Definition 1.1.1: A *statement* is a sentence which is either true or false; but not both.

A statement is also called a *proposition*. Statements will be denoted by letters such as p, q, r, etc. If p is the statement "I am happy" we write "p: I am happy".

Example 1

p : All men are mortal

q : Socrates is a man

$r: 2 + 2 = 5$

p, q, and r are statements.

Example 2

s: Come here! (s is not a statement).
Questions and commands are not statements.

Definition 1.1.2: The truthfulness or falsity of a statement is called its *truth value*.

Let T denote true and F denote false. In the examples above the statements p and q have truth value T. The statement r has truth value F.

Definition 1.1.3: A *paradox* is a sentence that can neither be true nor false.

Example 3

Let v be the sentence: "This statement is false." Then v is not a proposition.
Reason: If we say that the sentence is true, then the sentence itself asserts that it is false. If we say that it is false, then we must deduce it is true. So the sentence can neither be true nor false. It is a paradox or self-contradicting sentence.

Example 4: The sentence "There are no absolute truths" contradicts itself. It is a paradox.

Definition 1.1.4: A statement which refers to a single concept is called a *single statement* or a *propositional variable*.

Definition 1.1.5: A *compound* (or composite) statement is composed of single sub-statements and connectives.

The truth value of a compound statement is completely determined by its sub-statements and connectives. We will look now at five connectives.

1.2. CONNECTIVES

Definition 1.2.1: A *truth table* is a table which gives the truth values of the proposition under all possible assignments of its variables.

A truth table can be used to define the basic concepts in Logic.

(i) CONJUGATION

The conjugation of two statements p and q, is the compound statement 'p and q'; and is denoted by $p \wedge q$.

Example 5

If p: I like cake and q: I like ice-cream, then $p \wedge q$ is the statement "I like cake and ice-cream."

The truth table at the top of page 3 defines the conjunction $p \wedge q$.

Deduction: The *conjunction* of two statements is true if each component is true. This is the only case when the conjunction is true.

p	q	$p \wedge q$
T	T	T
T	F	F
F	T	F
F	F	F

Conjunction can be used to define the intersection of two sets: $A \cap B = \{x : x \varepsilon A \wedge x \varepsilon B\}$. We will study intersection of sets in chapter 2.

(ii) DISJUNCTION

The disjunction of two statements p, q is the compound statement p or q; and is denoted by $p \vee q$.

Example 6: If p: I like cake and q: I like ice-cream, then $p \vee q$ is the statement "I like cake or ice-cream."

Disjunction functions like the union of sets: $A \cup B = \{x : x \varepsilon A \vee x \varepsilon B\}$. The 'or' is an inclusive 'or.' In English 'or' means 'one or the other but not both.' For example, an air hostess asks: "Do you want coffee or tea?"—You can't say "both". But in Mathematics 'or' is inclusive, meaning 'either one or the other or both.'

The following truth table defines the disjunction $p \vee q$:

p	q	$p \vee q$
T	T	T
T	F	T
F	T	T
F	F	F

Deduction: The *disjunction* of two statements is false only if each component is false.

(iii) NEGATION

Given a statement p, its negation is the statement 'not p'; and is denoted $\sim p$.

Example 7

If p: I like cake, then $\sim p$: I do not like cake.

Note that you cannot simply put a 'not' if front of the statement and say "not I like cake." We do not speak like that. You can however say, "It is not the case that I like cake."

Negation is a unary connection, i.e. it operates on a single statement. Conjugation and disjunction are binary connectives uniting two statements. The table below defines $\sim p$:

p	$\sim p$
T	F
F	T

Deduction: A statement B is a *negation* of statement A if wherever A is true, B is false and wherever A is false, B is true.

To negate a statement, we cannot just put 'not' in front of it. We must write a proper English sentence.

Example 8: Let p: some men are smart. What is the negation of p?

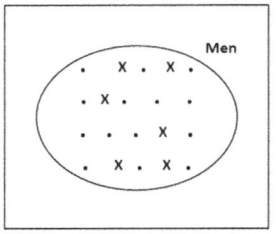

Some men are smart

Diagram 1
x represents a smart man

If you said "Some men are not smart," you are wrong. $\sim p$ is the statement: "It is not the case that some men are smart," but what does this mean precisely in ordinary English? We must understand the meaning of the word 'some'. 'Some' means 'there exists at least one.' Its negation means 'there are none.' Thus the negation of "Some men are smart" is "There are no smart men." Diagrams 1 and 2 illustrate this. We will come back to negating statements in Section 1.6.

Negation can be useful in proofs. A proof of any statement amounts to a disproof of its negation. To prove p is true, we can show that $\sim p$ is false. We will study proofs in Section 1.7.

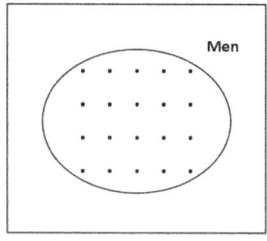

Diagram 2

There are no smart men

We can use simple statements and more than one connective to form larger compound statements. Their truth values are determined by the truth values of their simpler statements and the connectives used.

Example 9: The truth tables for $(p \wedge q) \vee \sim p$ and $q \vee \sim p$ are given below:

p	q	$p \wedge q$	$\sim p$	$(p \wedge q) \vee \sim p$	$q \vee \sim p$
T	T	T	F	T	T
T	F	F	F	F	F
F	T	F	T	T	T
F	F	F	T	T	T

Exercise: Construct the truth table for the compound statement $p \wedge (\sim q)$.

LOGICAL EQUIVALENCE

Definition 1.2.2: Two compound statements, each composed from the same simple statements, are said to be *logically equivalent* if they have the same truth values for all possible choices of the truth values for their simple statements. (The last columns of their truth tables are the same).

Logical equivalence is denoted by the symbol $\equiv$.

Note that the columns for $(p \wedge q) \vee \sim p$ and $q \vee \sim p$ in the above table are exactly the same. The two statements are therefore logically equivalent. We write $q \vee \sim p \equiv (p \wedge q) \vee \sim p$. If, for example, p: I like cake and q: I like ice-cream, this means that saying "I like ice-cream or I don't like cake" is equivalent to saying "I like cake and ice-cream or I don't like cake."

(iv) CONDITIONAL STATEMENTS

The conditional statement 'if p (is true), then q (is true)' or 'p implies q' is written $p \to q$ or $p \Rightarrow q$.

Example 10

A student says, "If I pass my exam then I will take the entire class out to dinner." Let p: I pass my exam and q: I take the entire class out to dinner. The student's statement is $p \to q$.

The following table defines the conditional $p \to q$:

p	q	$p \to q$
T	T	T
T	F	F
F	T	T
F	F	T

Deduction: The *conditional* $p \to q$ is true unless p is true and q is false.

Note that if the student in example 10 fails his exam and takes the class out to dinner he has not told a lie. We also see from the 2nd row of the table that a true statement cannot imply a false one. The last two rows show that beginning with lies we can prove anything.

The statement $p \to q$ is also read 'p is sufficient for q.'

Definition 1.2.3: Condition A is *sufficient* for condition B if B always occurs when A occurs.

In Example 10, the condition "Passing my exam" is sufficient for "taking the entire class to dinner." We can also write the student's statement as "I will take the entire class out to dinner, if I pass my exam" or 'q if p.' p is the antecedent and q is the consequent.

Exercise: Show that $p \to q \equiv (\sim p) \vee q$.

(v) BI-CONDITIONAL

The bi-conditional 'if and only if' is represented by the symbol '$\Leftrightarrow$' or '$\leftrightarrow$'. '$p \Leftrightarrow q$' means 'if p then q and if q then p'; or 'p implies q and q implies p.' The abbreviation 'iff' is sometimes used for 'if and only if.'

Example 11

Suppose that in example 10 the student says instead, "I will take the entire class out to dinner, if and only if I pass my exam." Let p: I take the entire class out to dinner and q: I pass my exam. Then the student's statement can now be written: $p \Leftrightarrow q$.

The following table defines the bi-conditional $p \Leftrightarrow q$:

p	q	$p \Leftrightarrow q$
T	T	T
T	F	F
F	T	F
F	F	T

Deduction: If p and q have the same truth value, then $p \Leftrightarrow q$ is true. If p and q have opposite truth values then $p \Leftrightarrow q$ is false.

Exercise

Verify that $(p \rightarrow q) \wedge (q \rightarrow p) \equiv p \Leftrightarrow q$.

This completes our study of the connectives $\wedge, \vee, \sim, \Rightarrow, \leftrightarrow$. We move on, now, to two important variations of the conditional statement.

1.3 CONVERSE AND CONTRAPOSITIVE

(i) CONVERSE

Definition 1.3.1: The *converse* of the implication $p \rightarrow q$ is the implication $q \rightarrow p$.

Example 12

Let p: It is a monkey, q: It has a tail, and $p \rightarrow q$: If it is a monkey then it has a tail. Then the converse $q \rightarrow p$: If it has a tail then it is a monkey.

Example 13

Show that $p \rightarrow q \not\equiv q \rightarrow p$.

Solution: We observe from the truth table that $p \to q$ and $q \to p$ do not have the same truth values for all possible choices of the truth values of their simple statements. Therefore, they are not logically equivalent.

p	q	$p \to q$	$q \to p$
T	T	T	T
T	F	F	T
F	T	T	F
F	F	T	T

A statement and its converse are therefore not equivalent. It is crucial to understand the difference. The truth of a statement does not imply the truth of its converse. If you are asked to prove an implication and you prove its converse instead, no marks will be awarded. You must be careful here as this is a common mistake.

Example 14

Prove that if $y = z$ then $x + y = x + z$ for any real number of x.

"Proof": $x + y = x + z$
$\Rightarrow -x + x + y = -x + x + z$
$\Rightarrow y = z$. □

You will get zero marks for this answer since you have proven the converse of what you were asked to prove. To prove an implication, you must start by assuming that the antecedent is true; and then show that the consequent follows.

(ii) CONTRAPOSITIVE

Definition 1.3.2: The *contrapositive* of an implication $p \to q$ is the implication $(\sim q) \to (\sim p)$; read 'if not q, then not p.'

Example 15

The contrapositive of the implication "If it is a monkey, then it has a tail" is the implication "If it does not have a tail, then it is not a monkey."

Example 16

Show that $p \to q \equiv \sim q \to \sim p$.

ELEMENTARY LOGIC

p	q	$\sim q$	$\sim p$	$\sim q \to \sim p$	$p \to q$
T	T	F	F	T	T
T	F	T	F	F	F
F	T	F	T	T	T
F	F	T	T	T	T

Since they always have the same truth values an implication and its contrapositive are equivalent.

To prove an implication, you can therefore prove its contrapositive. For example; to prove that if $x+2=4$ then $x=2$, you can show that if $x \neq 2$, then $x+2 \neq 4$.

1.4 NECESSARY AND SUFFICIENT CONDITIONS

We saw in Example 15 that saying "if it is a monkey, then it has a tail" is equivalent to saying that "if it does not have a tail, then it is not a monkey." In other words, "having a tail is necessary for being a monkey." The Venn diagram illustrates these statements. You will learn how to use Venn diagrams in Chapter 2 but we now have another way of expressing $p \to q$. We can say that q is *necessary* for p.

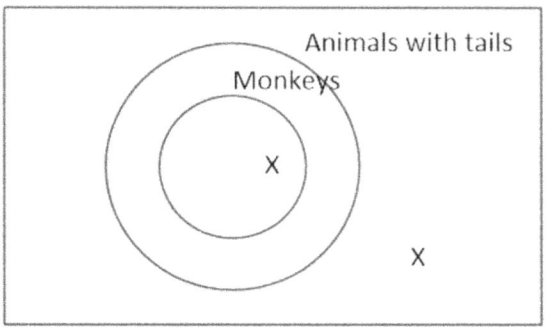

Diagram 3

Definition 1.4.1: Condition B is a *necessary* condition for condition A if A cannot occur unless B also occurs; i.e. *A only if B*.

Saying that "having a tail is necessary for being a monkey" is also equivalent to saying that "it is a monkey only if it has a tail," i.e. *p* only if *q*. So the symbol $\to$ can be read 'only if.'

9

FOUNDATIONS OF MATHEMATICS

Definition 1.4.2: If A occurs if and only if B occurs, then A is said to be a *necessary and sufficient* condition for B; i.e. 'A iff B' or $A \Leftrightarrow B$.

Exercise: Show that $A \Leftrightarrow B \equiv B \Leftrightarrow A$.

So, $A \to B$ means 'A is sufficient for B' or 'B is necessary for A' and $B \to A$ means 'A is necessary for B' or 'B is sufficient for A.' When we have both $A \to B$ and $B \to A$, i.e. $A \leftrightarrow B$, we say that 'A is necessary and sufficient for B' or 'B is necessary and sufficient for A.'

We recap now the many ways of translating $p \to q$:

1. If p, then q.
2. q if p.
3. p implies q.
4. p is sufficient q.
5. p only if q.
6. q is necessary for p.

Example 17

Let p: It is a man, q: It is mortal, then $p \to q$ can be written:
1. If it is a man, then it is mortal.
2. It is mortal if it is a man.
3. It is a man implies it is mortal.
4. It is a man is sufficient for it being mortal.
5. It is a man only if it is mortal.
6. Being mortal is necessary for being a man.

Diagram 4 illustrates these statements. They are equivalent to saying: "All men are mortals."

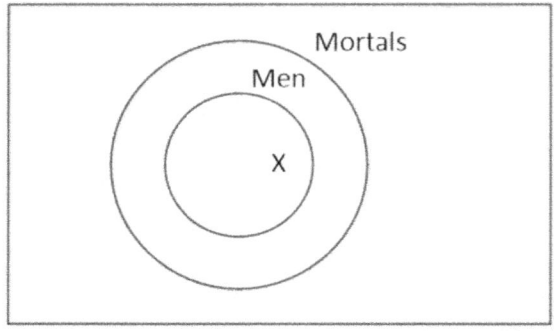

Diagram 4

ELEMENTARY LOGIC

Exercise

Write the statement "All mathematicians are smart" as a conditional statement and give five different ways of expressing it.

Exercise

Let p be the statement "It is sunny" and let q be the statement "It is hot." Write the following as conditional statements involving p and q:

   (i) It is sunny only when it is hot.
   (ii) If it is sunny, then it is hot.
   (iii) If it is hot, it is sunny.
   (iv) It is hot only when it is sunny.
   (v) It is sunny when and only when it is hot.

Exercise: Show that $A \to B \equiv\, \sim (A\, \wedge \sim B)$.

This formula will be useful for negating implications. Note that the right hand side, can be read 'It cannot happen that A occurs and B does not;' that is, 'B is necessary for A.'

1.5 TAUTOLOGIES AND CONTRADICTIONS

Definition 1.5.1: A compound statement which is true for all possible truth values of its component statements is called a *tautology*.

Example 18: Let p be the statement "Solomon is wise." p can be true or false depending on who Solomon is. The statement $p\, \vee \sim p$ which reads, "Solomon is wise or Solomon is not wise" is, however, always true. It is therefore a tautology.

p	$\sim p$	$p\, \vee \sim p$
T	F	T
F	T	T

The column in the truth table corresponding to the statement $p\, \vee \sim p$ contains only T's. This is the case for any statement which is a tautology.

Definition 1.5.2: A compound statement is called a *contradiction* if it is false for all possible truth values of its component statements.

Example 18 (continued): The statement $p \wedge \sim p$ reads "Solomon is wise and Solomon is unwise." It is always false; it is therefore a contradiction.

p	$\sim p$	$p \wedge \sim p$
T	F	F
F	T	F

The truth table will have only F's in the column corresponding to a contradiction.

Exercise: Show that the statement $(p \rightarrow q) \leftrightarrow (\sim q \rightarrow \sim p)$ is a tautology.

Tautologies make the laws of logic. Here are two of them:

(i) Law of Contra-position

$(p \rightarrow q) \leftrightarrow (\sim q \rightarrow \sim p)$.

(ii) Law of Syllogism

$(p \rightarrow q) \wedge (q \rightarrow r) \rightarrow (p \rightarrow r)$. (*Exercise:* Draw up the truth table. There will be 8 rows.)

1.6 QUANTIFIERS

Definition 1.6.1: $P(x)$ is a *propositional function* on a set A if $P(x)$ becomes a statement whenever any element $a \in A$ is substituted for x. $P(x)$ is also called an *open sentence*.

The set of elements $a \in A$ for which $P(a)$ is true is called the *truth set* of $P(x)$, denoted T_P.

Example 19

Let $P(x): x + 2 < 9$, where x is a real number. If $x = 5$, say, then $P(x)$ is true, if $x = 12$, say, then $P(x)$ is false. $P(x)$ is a propositional function.

Example 20

If A is the set of real numbers, $\mathbb{R}$, then the truth set of $x + 2 < 9$ is $T_P = \{x : x < 7\}$.

Since T_p is non empty, we can say "there exists an x such that $x + 2 < 9$," and write $\exists x : x + 2 < 9$ or $\exists x, P(x)$. The symbol $\exists$ is called the *existential quantifier.*

Exercise: Find the truth set for $P(x): x+2 < x+3$

In the above exercise, the solution is the set of all x. We say $P(x)$ is true for all x; and write $\forall x, P(x)$. The symbol $\forall$ is called the *universal quantifier.*

For $P(x): x+2 < 9$ to be true x must be less than 7, so it is not true that for all x, $P(x)$. We write $\sim(\forall x, P(x))$ or $\exists x, \sim P(x)$, meaning there exists an x such that $P(x)$ is not true. So the negation of $\forall x, P(x)$ is $\exists x, \sim P(x)$.

If a statement $q(x)$ is never true, we write $\sim(\exists x, q(x))$ or $\forall x, \sim q(x)$. The negation of $\exists x, q(x)$ is therefore $\forall x, \sim q(x)$.

Example 21

Translate the statement "Some men are smart" into the language of logic, then write it's negation.

Solution: Let $P(x)$ denote "man x is smart." "Some men are smart" can be written: $\exists x, P(x)$, i.e. "There is at least one x for which $P(x)$ is true."

Its negation is: $\sim(\exists x, P(x))$ or 'for all x, $P(x)$ is not true,' written: $\forall x, \sim P(x)$.

In words: "All men are non-smart" or "All men are stupid."
So the negation of "Some men are smart" is "All men are stupid."

Example 22

What is the negation of "All men are stupid"?

Solution: Let $q(x)$: man x is stupid.
Given statement: $\forall x, q(x)$.
Negation: $\sim(\forall x, q(x))$ or $\exists x, \sim q(x)$.
In words: "There exists at least one man who is not stupid" or "Some men are smart."

1.7 ARGUMENTS

Definition 1.7.1: A statement used to arrive at a conclusion is called as a *premise*.

We want to know that if all the premises used in the argument are true, then the conclusion must necessarily be true.

Definition 1.7.2: An argument is *valid,* iff the conclusion is true whenever the premises are true.

Example 23: Consider the following argument:

Premise 1: I am either a man or a woman

Premise 2: I am not a woman

Conclusion: I am a man

Is this a valid argument?

Solution: Let A: "I am a man" and B: "I am a woman." The argument can be written:

Premise 1: $A \vee B$

Premise 2: $\sim B$

Conclusion: A

We can determine the validity of the argument by constructing a truth table:

C		P_2	P_1
A	B	~B	$A \vee B$
T	T	F	T
T	F	T	T
F	T	F	T
F	F	T	F

The truth table shows that whenever both the premises are true, the conclusion is true. Therefore the argument is valid. □

Note: If one of the premises in a logically valid argument is false, it is possible to arrive at a conclusion which is false. The next example illustrates this.

Example 24

P_1: Either I am poor or I am rich

P_2: I am not poor

C: I am rich

The argument is valid but the conclusion is not true. I am certainly not rich! This is because P_1 is false. In logic, we are testing the validity of arguments and are only interested in when the premises are true.

ELEMENTARY LOGIC

Example 25: Test the validity of the argument

P_1: I am either a man or a woman

P_2: I am a man

C: I am not a woman

Solution: Let A: "I am a man" and B: "I am a woman." The argument can be written:

P_1: $A \vee B$

P_2: A

C: $\sim B$

P_2		C	P_1
A	B	$\sim B$	$A \vee B$
T	T	F	T
T	F	T	T
F	T	F	T
F	F	T	F

There is a case when the premises are true and conclusion is false. The argument is therefore invalid. (To make it valid you would need to specify "I am a woman or a man but not both," then P_1 is no longer $A \vee B$).

We need not construct truth tables each time we have to check the validity of an argument. We can instead try to recognise the form of argument. We go on now to some classical valid forms.

SOME CLASSICAL ARGUMENTS

1. Modus Ponens

$P_1 : A \rightarrow B$

$P_2 : A$

$C : B$

The truth table below shows that whenever the premises are true the conclusion is true. This form of argument is called *Modus Ponens*, from the Latin *'modus'* (way) and *'ponere'* (to put). If $A \rightarrow B$ and we 'put' A, then we have B.

P_2	C	P_1
A	B	$A \to B$
T	T	T
T	F	F
F	T	T
F	F	T

Example 26

Test the validity of the argument:

P_1: If Plato is a man then he is fallible

P_2: Plato is a man

C: Plato is fallible

Solution: First convert to the language of logic:

Let A: "Plato is a man" and let B: "He is fallible." The given argument can be written:

$P_1 : A \to B$

$P_2 : A$

$C : B$

We can recognise it as a Modus Ponens argument. Therefore the argument is valid.

2. Modus Tollens

$P_1 : A \to B$

$P_2 : \sim B$

$C : \sim A$

The truth table below shows that whenever the premises are true the conclusion is true. *Modus Tollens* is a classically valid argument. The name comes from the Latin 'modus' (way) and 'tollere' (to take). If $A \to B$ is true and we 'take away' B then we do not have A.

A Modus Tollens argument essentially illustrates that an implication and its contrapositive are equivalent.

		P_1	P_2	C
A	B	$A \to B$	$\sim B$	$\sim A$
T	T	T	F	F
T	F	F	T	F
F	T	T	F	T
F	F	T	T	T

Example 27: We can recognise the following as a valid Modus Tollens argument:

P_1: If it is sunny then I am at the beach

P_2: I am not at the beach

C: It is not sunny

Fallacies resembling Modus Ponens and Modus Tollens

(i) *Fallacy of Affirming the Consequent.* Consider the following argument:

$P_1 : A \to B$

$P_2 : B$

$C : A$

A	B	$A \to B$
T	T	T
T	F	F
F	T	T
F	F	T

The truth table shows that there is a case when the premises are true, but the conclusion is false. The argument is therefore invalid. It is known as the *Fallacy of Affirming the Consequent.*

(ii) *Fallacy of Denying the Antecedent.* Consider the following argument:

$P_1 : A \to B$

$P_2 : \sim A$

$C : \sim B$

A	B	A → B	~B	~A
T	T	T	F	F
T	F	F	T	F
F	T	T	F	T
F	F	T	T	T

The truth table shows that it is invalid. It is known as the *Fallacy of Denying the Antecedent*.

Exercise: Identify the form of the following arguments:

(i) P_1: If you are happy then you are smiling

P_2: You are smiling

C: You are happy

(ii) P_1: If you are happy then you are smiling

P_2: You are not happy

C: You are not smiling

3. Hypothetical Syllogism

Example 28: Consider the following argument

P_1: Socrates is a man

P_2: All men are mortal

C: Socrates is mortal

Let A: "It is Socrates," B: "It is a man" and C: "It is mortal." The given argument can be written:

P_1: $A \to B$

P_2: $B \to C$

C: $A \to C$

A truth table will verify that this is a valid argument. It is called *hypothetical syllogism* and is due to Aristotle.

Fallacies Resembling Syllogism

(i) P_1: $B \to A$

P_2: $C \to A$

C: $B \to C$

(ii) P_1: $A \to B$

P_2: $A \to C$

C: $B \to C$

Exercise:

Verify, using truth tables, that the above arguments are invalid. Also, for each, construct a corresponding practical example.

4. Disjunctive Syllogism (See Example 23)

P_1: $A \vee B$

P_2: $\sim B$

C: A

5. Constructive Dilemma

P_1: $(A \to B) \wedge (C \to D)$

P_2: $A \vee C$

C: $B \vee D$

Exercise

Complete the truth table for the *Constructive Dilemma*. There are 16 rows. You will find that whenever the premises are true, the conclusion is true.

A	B	C	D	$A \to B$	$C \to D$	$(A \to B) \wedge (C \to D)$	$A \vee C$	$B \vee D$
T	T	T	T	T	T	T	T	T
T	T	T	F	T	F	F	T	T
⋮	⋮	⋮	⋮	⋮	⋮	⋮	⋮	⋮

A constructive dilemma is fundamentally a combination of two *Modus Ponens* arguments. The following is an example:

Example 29

>P_1: If it rains I will stay inside but if it shines I will go out
>
>P_2: It will either rain or shine
>
>C: I will either stay inside or go outside

6. Destructive Dilemma

Destructive dilemma is fundamentally a combination of two *Modus Tollens* arguments:

>P_1: $(A \rightarrow B) \wedge (C \rightarrow D)$
>
>P_2: $\sim B \vee \sim D$
>
>C: $\sim A \vee \sim C$

Example 30

>P_1: If I take Math I will work problems and if I take English I will write essays
>
>P_2: I will not work problems or I will not write essays
>
>C: Either I will not take Math or I will not take English

Let A: I take Math, B: I will work problems, C: I take English, and D: I will write essays. The argument can be written in the form above, thus poses a destructive or negative dilemma.

1.8 PROOF BY CONTRADICTION

A truth cannot imply a falsity. So if the conclusion in a logically valid argument is false, then, one of the premises must be false. If there are only two premises and one of them is known to be true, then the other must be false. The negation of the false premise is therefore true.

This is the idea behind proof by contradiction. We can prove the truth of a statement by disproving its negation. To do this, we must show that the negation when used with other premises which we know to be true, gives (through a logically valid argument) a conclusion which is false.

Proof by Contradiction is also called *Reductio ad Absurdum*. It is essentially reducing an argument to an absurdity. Here is the structure:

Proof (by contradiction)

To prove A (is true). We know $A \vee \sim A$ is true (tautology). So if we know that $\sim B$ is true (i.e. B is false), we only need show that $\sim A \to B$ is true. Then $\sim A$ must be false. Therefore A is true.

A proof by contradiction is essentially a *Modus Tollens* argument:

$P_1 : \sim A \to B$

$\underline{P_2 : \sim B}$

$C : \sim (\sim A)$

Example 31

Prove statement A, where A: "All Greeks are mortal."

(To show that A is true, we show that $\sim A$ is false. To show that $\sim A$ is false, we show that $\sim A \to B$, where we know that $\sim B$ is true, or B is false. So choose $\sim B$ as a statement known to be true. Let $\sim B$: "All humans are mortal.")

Proof. Let us assume that all humans are mortal.
Some Greeks are not mortal implies that some humans are not mortal.

But all humans are mortal (contradiction).

∴ All Greeks are mortal. □

Example 32

Suppose that whenever it is raining it is wet. Suppose that it is now raining. Prove, by contradiction, that it is wet.

Proof (by Contradiction):

Let us assume that it is raining. Let A: "It is wet" and $\sim B$: "It is raining." We show that $\sim A \to B$ where $\sim B$ is true. Therefore A is true:

It is not wet implies that it is not raining.

But it is raining (contradiction).

Therefore it is wet. □

We used simple examples above to illustrate the form of the proof. There was hardly anything to prove. Usually, showing that $\sim A \to B$ involves some work. Here is a more challenging example.

Example 33: Prove that if x^2 is an odd integer, then x is odd.

Proof (by contradiction):

Let *A: x is odd* and *~B: x^2 is an odd integer*. We show that $\sim A \rightarrow B$:

x is even $\rightarrow x = 2a$, where $a \in \mathbb{Z}$.
$\rightarrow x^2 = 4a^2 = 2(2a^2) = 2k$, where $k \in \mathbb{Z}$.
$\rightarrow x^2$ is even (*B*).
But x^2 is odd (~*B* is true).
∴ x cannot be even.
∴ x is odd. □

Disproof by Counter Example

If a statement is true, then the statement holds in every instance in which it claims to hold. We can, therefore, disprove a statement by finding a single instance in which it claims to hold, but does not. Such an instance is called a *counter example*. Therefore a single counter example disproves a statement.

Note: An example does not prove a general statement, but a counter example disproves it.

Example 34: To disprove the statement "Nobody knows the trouble I've seen."

Come up with a counter example— "My mother knows all my troubles." □

◇◇◇

1.9 FEATURE MATHEMATICIAN

ARISTOTLE

THE FOUNDER OF LOGIC

Plato (left) and Aristotle (right), a detail of *The School of Athens*, a fresco by Raphael, 1509-11.

Aristotle (384-322B.C.) was born in Macedonia, Greece. At this time Greece was the centre of the world and the Greeks took great pride in knowledge. This was the heroic age of Mathematics that produced philosophers like Socrates, Plato, and Aristotle, who number among the greatest thinkers of all time.

Influenced by Pythagoras, Plato was convinced that geometry was the key to unlocking the secrets of the universe and established a prestigious school in Athens. On the door of Plato's academy was inscribed the motto: "Let no one ignorant of Geometry enter here." The front cover image of this book is a portrayal of *The School of Athens* as depicted by Renaissance artist, Raphael. At the very center stand Plato and Aristotle carrying their books in their left hands. They are surrounded by all the great philosophers of antiquity. Pythagoras is seated at lower left. You can read about the other figures in the painting online.

Aristotle was the greatest and most influential of Plato's students. His logical works contain the earliest formal study of logic that we now have. He was the first to demonstrate the principles of reasoning, by employing variables to show the underlying logical form of an argument; as we have done in this chapter. The law of Syllogism is due to him. He was the first to deal with the principles of contradiction; and to prove invalidity by constructing counterexamples. The work of George Boole and Augustus De Morgan in the first half of the nineteenth century then initiated the study of modern logic. Aristotle was also a brilliant scientist and made contributions to almost every subject including physics, biology and other natural sciences. He tutored Alexander, the Great and other rulers.

Over 2,300 years ago Aristotle wrote: "*We are what we repeatedly do. Excellence, then, is not an act, but a habit.*" It is still so applicable to students today.

1.10 CHAPTER EXERCISES

1. If p, q and r represent true-false statements, show that:
(i) $\sim (p \vee q) \equiv (\sim p) \wedge (\sim q)$
(ii) $\sim (p \wedge q) \equiv \sim p \vee \sim q$
(iii) $(p \to q) \to r$ is *not* logically equivalent to $(p \to r) \vee (q \to r)$

2. Determine which of the following statements is a tautology or a contradiction:
(i) $[(p \to q) \wedge p] \to q$
(ii) $(p \to \sim q) \wedge (p \wedge q)$
(iii) $(\sim p \vee q) \vee (p \wedge \sim q)$
(iv) $(p \to q) \wedge (q \to r) \to (p \to r)$

3. Determine whether or not the following are valid arguments:
(i) Borogoves are mimsy whenever it is brillig
   It is now brillig, and this thing is a borogove
   ———
   Hence this thing is mimsy

(ii) If I go to sleep I will not finish my work and If I stay up I will be tired
   I will finish my work or I will not be tired
   ———
   I will not go to sleep or I will not stay up

4. Identify the type of argument used in each of the following:
(i) If it rains I will not go to the beach
   It is raining
   ———
   I am not going to the beach

(ii) If it rains I will not go to the beach
   I am going to the beach
   ———
   It is not raining

5. In the Land of Oz people either always speak the truth or lies. A traveler met a man named Mr. Dilemma and asked him if he was a liar or a truth-teller. Mr. Dilemma mumbled so softly the traveler did not hear him, so he asked Mr. Logic, who was standing right next to him, what Mr. Dilemma said. Mr. Logic replied, "He said that he was a liar." Is Mr. Logic a liar or a truth-teller?

6a. State the converse and contra-positive of the following statement:

"All men are mortal"

b. Determine whether or not the two statements are logically equivalent:

"If I go you will go"

"If you do not go, I will not go"

c. In the following statement determine whether one condition is necessary or sufficient for the other.

"I will come if it does not rain"

d. Convert the following into an "if ... then" statement:

"It is necessary for you to eat in order to live"

e. Write the negation of the following statement:

"If you can sing then sing."

f. Write the negation of the statement:

"All men are mortal."

7. Determine whether or not the following argument is a valid proof by contradiction:
If 1 is more than 2, then 2 is more than 3. But 2 is less than 3, hence 1 is less than 2.

8. If $p \to q$ is a tautology, we say that p *logically implies q.*
a. Show that:
   (i) $(p \to q) \wedge (q \to r)$ logically implies $(p \to r)$
   (ii) If $p \wedge \sim q$ is a contradiction then p logically implies q.
b. Decide whether or not:
   (i) p logically implies $p \wedge q$
   (ii) p logically implies $p \vee q$

CHAPTER REFERENCES

Herbert B. Enderton, *A Mathematical Introduction to Logic*, 2 ed. (New York: Academic/Elsevier, 2006).

Michael C. Gemignani, *Basic Concepts of Math and Logic* (Mass: Addison-Wesley Publishing Co., Inc., 1968).

P.A. Morris, *Introduction to Algebra*, 3rd edition (Department of Mathematics, UWI, St Augustine, 1996).

CHAPTER TWO

SETS

In this chapter we look at some basic ideas of Set Theory; such as, the empty set, subsets of sets, universal sets, Venn diagrams and the operations of intersection, union, complements and set difference. The emphasis will be on learning the language of sets, and the use on proper notations. We discuss how to establish equality of sets, both formally and by using set algebra. We then apply Set Theory to proving the validity of arguments.

2.1 BASIC IDEAS AND NOTATIONS

Intuitively a set is a collection of objects. The objects are called the *elements* or *members* of the set. Sets are usually denoted by capital letters. We write '$x \in S$' to say that 'x is an element of the set S' or 'x belongs to S'; and $y \notin S$ for 'y is not a member of S.' We can use Venn Diagrams to represent sets:

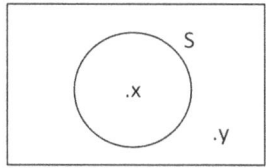

We use curly brackets when listing or describing the elements of a set; for example, $B = \{x : x \text{ is even}\}$ reads B is the set of all elements x, such that x is even.

We build our set theory from our definitions and axioms. Axioms are intuitive truths or assumptions which we accept as being true.

Axiom of Extension:

Two sets are equal iff they both have the same members.

Axiom of Empty Set: There is a set having no members.

Definition 2.1.1: The *empty* (null) set is the set which contains no elements.

The empty set is denoted $\{\}$ or $\emptyset$, with the symbol $\emptyset$ being preferred. Note that $\{\emptyset\} \neq \emptyset$ (after all, a man with an empty container is better off than a man with nothing). The empty set is unique, that is, there is only one empty set. For example, $\{x : x \neq x\} = \emptyset$ and the set of students in this class under 5 years old $= \emptyset$, but they are the same empty set. This empty set is a very interesting set with very interesting properties, for example, $\emptyset \in \{\emptyset\}$ but $\emptyset \notin \emptyset$.

SUBSET

Definition 2.1.2: A is a *subset* of B iff $x \in A$ implies $x \in B$. Symbolically, we write $A \subseteq B$.

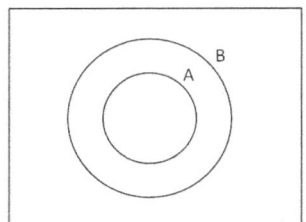

A is a subset of B

Theorem 2.1.1: The empty set is a subset of every set.

Proof: By definition 2.1.2, we have nothing to prove since there is no x in $\emptyset$. We say that the statement is vacuously true.

We may also prove the statement by contradiction:

Suppose that the empty set is not a subset of every set. Then there exists a set $A : \emptyset \nsubseteq A$.
$\Rightarrow \exists x \in \emptyset : x \notin A$. This is a contradiction since the empty set has no elements.
$\therefore$ The empty set is a subset of every set. □

Theorem 2.1.2: Every set is a subset of itself.

Proof: The proof is easy if the set is non-empty as $x \in A \Rightarrow x \in A$, $\therefore A \subseteq A$.

If $A = \emptyset$, then by theorem 2.1.1, we have nothing further to prove since the empty set is a subset of every set. $\therefore \emptyset \subseteq \emptyset$.

So, in either case, the set is a subset of itself. □

Exercise: Prove, by contradiction, that the empty set is a subset of itself.

We sometimes have to distinguish between a set and its subsets; so we make the following definition.

Definition 2.1.3: B is called a *proper subset* of A iff B is a subset of A, but B is not equal to A (i.e. $B \subseteq A$ and $B \neq A$).

Let us decide to use the symbol $\subseteq$ to denote subset and the symbol $\subset$ to denote proper subset. For example, $\{1\} \subset \{1, 2\}$ and $\{1\} \subseteq \{1, 2\}$. However, $\{1, 2\} \subseteq \{1, 2\}$, but $\{1, 2\} \subset \{1, 2\}$ is not true. We will not always make the distinction, however. If B is not a subset of A, we write: $B \nsubseteq A$.

We use the idea of subset to define equality of sets:

Definition 2.1.4: Two sets A and B are *equal* iff every element in A is in B and every element in B is in A, i.e.

$$A = B \text{ iff } A \subseteq B \text{ and } B \subseteq A.$$

Definition 2.1.4 is a key definition. We will use it to prove formally the equality of sets.

In any given situation all the sets under consideration are likely to be subsets of a bigger single set, called the universal set; denoted U. For example, if we are talking only about real numbers then $U = \mathbb{R}$. In Venn diagrams, we use rectangles to represent universal sets.

Note that if the elements of a set A are sets themselves, i.e. A is a set of sets, we call A 'a *family of sets*'.

Definition 2.1.5: The family of all subsets of a set A is called the *power set* of A, denoted by $\wp_A$. Thus $\wp_A = \{X : X \subseteq A\}$.

Example 1

(i) If $S = \{a, b\}$, then $\wp_S = \{\{a\}, \{b\}, \emptyset, \{a, b\}\}$

(ii) $\wp_\emptyset = \{\emptyset\}$ (iii) $\wp_{\{\emptyset\}} = \{\emptyset, \{\emptyset\}\}$.

2.2 OPERATIONS ON SETS

Definition 2.2.1: Let A and B be two sets. The *union* of A and B, denoted $A \cup B$, is the set of all elements which belong to A or to B.

Thus we have:
$$A \cup B = \{x : x \in A \text{ or } x \in B\}.$$

The 'or' here is the same as the logical connective '$\vee$'.

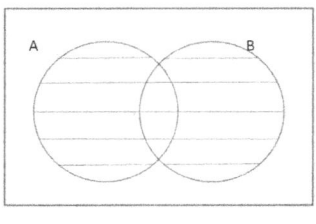

$A \cup B$

Definition 2.2.2: The *intersection* of two sets A and B, denoted $A \cap B$, is the set of all elements which belong to both A and to B.

Thus we have:
$$A \cap B = \{x : x \in A \text{ and } x \in B\}.$$

The 'and' here is the logical connective '$\wedge$'.

We may also write $A \cap B = \{x : x \in A \, , \, x \in B\}$. The 'comma' here means 'and.'

Note: (i) $A \cap B = B \cap A$ (ii) $A \cap B \subset A$ (iii) $A \cap B \subset B$ (iv) $A \subset A \cup B$
(v) $A \cup B \not\subset A$

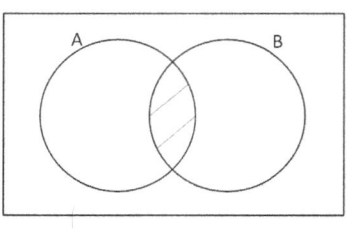

$A \cap B$

Definition 2.2.3: If two sets A and B have no elements in common, then we say that A and B are *disjoint*.

If A and B are disjoint then $A \cap B = \varnothing$.

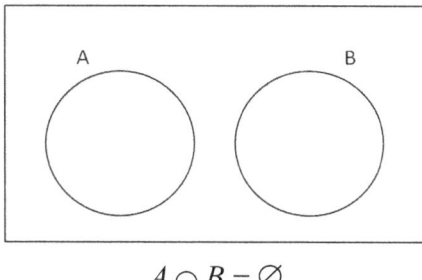

$A \cap B = \varnothing$

Definition 2.2.4: Given two sets A and B, their *difference*, denoted $A - B$, is the set of all elements which belong to A but which do not belong to B.

Thus we have:
$$A - B = \{x : x \in A \wedge x \notin B\}.$$

$A - B$ is also denoted by A/B and $A \sim B$.

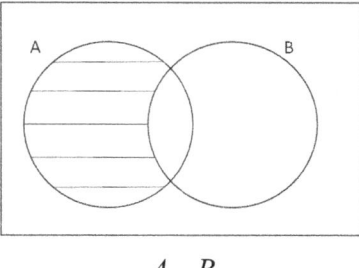

$A - B$

We can see from the Venn diagram that $A - B \subseteq A$ but $A - B \not\subseteq B$. Also $A - B \subseteq A \cup B$. You will be able to prove these theorems shortly.

The sets $A - B$, $A \cap B$ and $B - A$ are mutually disjoint; and there union is $A \cup B$.

Thus:
$$(A - B) \cup (A \cap B) \cup (B - A) = A \cup B.$$

This is one way of partitioning $A \cup B$.

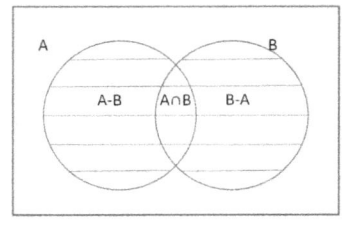

$$(A - B) \cup (A \cap B) \cup (B - A) = A \cup B$$

Definition 2.2.5: The *complement* of a set A is the set of all elements in U which do not belong to A.

The complement of A is denoted by A^c or A'. Thus:

$$A^c = \{x : x \notin A\}.$$

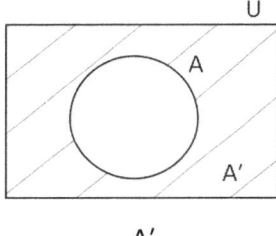

A'

Note: (i) $A \cup A^c = U$ (ii) $A \cap A^c = \emptyset$ (iii) $U^c = \emptyset$ (iv) $\emptyset^c = U$ (v) $(A')' = A$.

We have been illustrating sets with Venn diagrams but note that a Venn diagram is not a proof. It is only a pictorial representation. We now go on to formal proofs of set equalities.

2.3 FORMAL PROOF OF SET EQUALITIES

Some set theorems are easy to prove directly from the definitions given in section 2.2.

Example 2

Prove that for any two sets A and B

$$A - B = A \cap B'.$$

Proof (directly from set definitions)

$$A - B = \{x : x \in A, x \notin B\}.$$
$$= \{x : x \in A, x \in B'\}$$
$$= A \cap B'. \square$$

We, however, use Definition 2.1.4 in order to formally prove equality of sets:

$$A = B \text{ iff } A \subseteq B \text{ and } B \subseteq A.$$

We would need Definition 2.1.2 to first prove that one set is a subset of another:

$$A \subseteq B \text{ iff } x \in A \rightarrow x \in B.$$

We illustrate this algorithm using the same example:

Example 2 (continued): Prove that for any two sets A and B

$$A - B = A \cap B'.$$

Proof (from the definition of set equality):

[We must show that (i) $A - B \subseteq A \cap B'$ and (ii) $A \cap B' \subseteq A - B$.

(i) To show that $A - B \subseteq A \cap B'$, we must show that $x \in (A - B) \rightarrow x \in (A \cap B')$].

Let $x \in (A - B)$. Then $x \in A$ and $x \notin B$. Therefore, $x \in A$ and $x \in B'$. Therefore $x \in (A \cap B')$.

$\therefore A - B \subseteq A \cap B'$.

[(ii) To show that $A \cap B' \subseteq A - B$ we must show that $x \in (A \cap B') \rightarrow x \in (A - B)$].

Let $x \in (A \cap B')$. Then $x \in A$ and $x \in B'$. Therefore $x \in A$ and $x \notin B$. Therefore $x \in (A - B)$.

$\therefore A \cap B' \subseteq A - B$.

Since $A - B \subseteq A \cap B'$ and $A \cap B' \subseteq A - B$, we have $A - B = A \cap B'$. $\square$

Notice that part (ii) of the above proof was the reverse of part (i). In this case, we can write up parts (i) and (ii) of simultaneously:

Alternate Proof:

$x \in A - B \Leftrightarrow x \in A$ and $x \notin B \Leftrightarrow x \in A$ and $x \in B' \Leftrightarrow x \in A \cap B'$. $\therefore A - B = A \cap B'$. □

The double implication sign means that the proof holds in either direction. This method is not recommended for beginners, however.

We can also use logic to prove theorems on sets. There is, however, a difference between a set and a statement. To transform the given set theorem into the language of logic we first have to define variables $p : x \in A$ and $q : x \in B$. We then write the theorem as a logical statement and prove that it is a tautology. We will illustrate this technique in the next example.

Example 3

Prove that if A is a subset of B and B is a subset of C then A is a subset of C; that is:

$$A \subseteq B \wedge B \subseteq C \rightarrow A \subseteq C.$$

Proof: Assume that $A \subseteq B$ and $B \subseteq C$. We must show that $A \subseteq C$. Let $x \in A$.
$x \in A \Rightarrow x \in B$ (since $A \subseteq B$)

$\Rightarrow x \in C$ (since $B \subseteq C$)

$\therefore A \subseteq C$. □

Alternate proof using truth tables

Let $p : x \in A$, $q : x \in B$ and $r : x \in C$. Then $A \subseteq B \equiv p \rightarrow q$, $B \subseteq C \equiv q \rightarrow r$ and $A \subseteq C \equiv p \rightarrow r$. The given result can be converted into the language of logic. It becomes:

$$(p \rightarrow q) \wedge (q \rightarrow r) \rightarrow (p \rightarrow r).$$

We can draw the truth table or recognise it as the law of syllogism. This proves the result. □

DE MORGAN'S LAWS

The following inequalities are known as *De Morgan's Laws:*

(i) $(A \cap B)^c = A^c \cup B^c$ (ii) $(A \cup B)^c = A^c \cap B^c$

Proof of (i)

Let $x \in (A \cap B)^c$.

Then $x \notin (A \cap B)$

$\Rightarrow x \notin A$ or $x \notin B$

$\Rightarrow x \in A^c$ or $x \in B^c$

$\Rightarrow x \in (A^c \cup B^c)$.

Hence $(A \cap B)^c \subseteq A^c \cup B^c$. (1)

Conversely, let $x \in (A^c \cup B^c)$.

Then $x \in A^c$ or $x \in B^c$

$\Rightarrow x \notin A$ or $x \notin B$

$\Rightarrow x \notin (A \cap B)$

$\Rightarrow x \in (A \cap B)^c$.

Hence $A^c \cup B^c \subseteq (A \cap B)^c$. (2)

From (1) and (2) we have $(A \cap B)^c = A^c \cup B^c$. □

The above proof used the fact that $x \notin (A \cap B) \Leftrightarrow x \notin A$ or $x \notin B$. This needs some explanation. If I say, for example, "you are not 'tall and handsome,'" it means that either "you are not tall" or "you are not handsome" (the 'or' is inclusive of the case "you are not tall and you are not handsome").

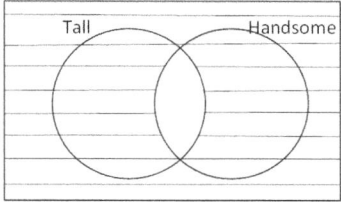

The shaded area represents those who are not tall and handsome

This corresponds to a statement which you were asked to prove at the end of Chapter 1:

$$\sim(p \wedge q) \equiv \sim p \vee \sim q.$$

De Morgan's Laws (continued)

(ii) $(A \cup B)^c = A^c \cap B^c$

Proof:

Let $x \in (A \cup B)^c$.

$x \in (A \cup B)^c \leftrightarrow x \notin (A \cup B) \leftrightarrow x \notin A \text{ and } x \notin B \leftrightarrow x \in A^c \text{ and } x \in B^c \leftrightarrow x \in (A^c \cap B^c)$.

$\therefore (A \cup B)^c = A^c \cap B^c$. □

Here again, the statement $x \notin (A \cup B) \leftrightarrow x \notin A$ and $x \notin B$ can do with some illustration. If I say "you are not taking the option 'Math or Computer Science,'" it means that "you are not taking Math and you are not taking Computer Science."

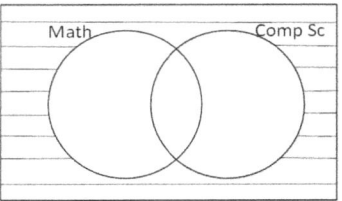

The shaded area represents those who are not taking Math or Comp Sc.

This corresponds to another statement which you were asked to prove at the end of Chapter 1:

$$\sim(p \vee q) \equiv (\sim p) \wedge (\sim q).$$

So we have two important results:

> (i) $x \notin (A \cap B) \leftrightarrow x \notin A \text{ or } x \notin B$
>
> (ii) $x \notin (A \cup B) \leftrightarrow x \notin A \text{ and } x \notin B$

The next section deals with yet another method of establishing results on sets.

2.4 SET ALGEBRA

The study of the operations $\cup, \cap, -, '$, together with the $\subseteq$ relation is called Set Algebra. The algebra of sets is similar to the algebra of real numbers with operations $+, -, \div, \times$. Here are some basic laws:

Theorem 2.4.1: For all sets A, B and C in a given universe:

(i) $A \cup \emptyset = A$, $A \cap \emptyset = \emptyset$, $A \cup U = U$, $A \cap U = A$, $A \cup A^c = U$, $A \cap A^c = \emptyset$.

(ii) $U^c = \emptyset$, $\emptyset^c = U$, $(A')' = A$.

(iii) $A \subseteq A \cup B$, $B \subseteq A \cup B$, $A \cap B \subseteq A$, $A \cap B \subseteq B$.

(iv) $A \cup B = B \cup A$, $A \cap B = B \cap A$ (Union and intersection are commutative).

(v) $A \cup (B \cup C) = (A \cup B) \cup C$ (Union is associative).

(vi) $A \cap (B \cap C) = (A \cap B) \cap C$ (Intersection is associative).

(vii) $A \cup (B \cap C) = (A \cup B) \cap (A \cup C)$ (Union distributes over intersection).

Also $(A \cap B) \cup C = (A \cup C) \cap (B \cup C)$ (Union distributes over intersection on the right).

(viii) $A \cap (B \cup C) = (A \cap B) \cup (A \cap C)$ (Intersection distributes over union).

Also $(A \cup B) \cap C = (A \cap C) \cup (B \cap C)$ (Right distributive).

(ix) $(A \cap B)^c = A^c \cup B^c$, $(A \cup B)^c = A^c \cap B^c$ (De Morgan's Laws).

(x) $A - B = A \cap B^c$.

(xi) $A \subseteq B \rightarrow A \cup C \subseteq B \cup C$, $A \subseteq B \rightarrow A \cap C \subseteq B \cap C$.

Also $A \subseteq B \rightarrow A \cap B = A$, $A \subseteq B \rightarrow A \cup B = B$.

You should be able to formally establish the proof of all these results and illustrate them with diagrams.

Exercise

Prove formally that if A is a subset of B then:

(i) $A \cup C \subseteq B \cup C$ (ii) $A \cap C \subseteq B \cap C$ (iii) $A \cap B = A$ (iv) $A \cup B = B$.

Also illustrate the results with Venn Diagrams.

We can use the above results along with other well established results to simply the algebra of sets; and prove further results.

Definition 2.4.1: The *Symmetric Difference* of A and B, denoted by $A \Delta B$, is defined to be the set of elements that belong either to A and not to B or to B and not to A. That is,

$$A \Delta B = (A - B) \cup (B - A).$$

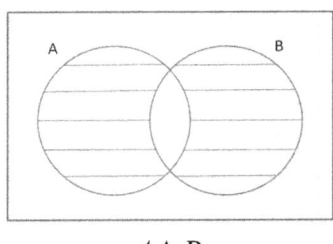

$A \Delta B$

Example 4

Use Set Algebra to establish the following results:

(i) $A \Delta A = \emptyset$ (ii) $A \Delta \emptyset = A$ (iii) $A \Delta U = A'$

Proof:

(i) $A \Delta A = (A - A) \cup (A - A)$ [By definition of Δ]

$= (A \cap A^c) \cup (A \cap A^c)$ [Theorem 2.4.1(x)]

$= \emptyset \cup \emptyset$ [Theorem 2.4.1(i)]

$= \emptyset . \square$

(ii) $A \Delta \emptyset = A$

Proof:

$A \Delta \emptyset = (A - \emptyset) \cup (\emptyset - A)$ [By definition]

$= (A \cap \emptyset^c) \cup (\emptyset \cap A^c)$ [Theorem 2.4.1(x)]

$= (A \cap U) \cup \emptyset$ [Theorem 2.4.1(i)]

$= A \cup \emptyset$ [Theorem 2.4.1(i)]

$= A . \square$

You should always state the theorems used in your proof. We omit them in the next part for the sake of brevity.

(iii) $A \Delta U = A'$

Proof:

$$A \Delta U = (A - U) \cup (U - A) = (A \cap U^c) \cup (U \cap A^c) = (A \cap \emptyset) \cup (U \cap A^c) = \emptyset \cup A^c = A^c. \square$$

Example 5

Let A, B and C are sets in a given universe.

(i) Prove formally that if $A \subseteq B$, then $C - B \subseteq C - A$

(ii) Simplify $[(A \cup B \cup C) \cap (A \cup B)] - [(A \cup (B - C)) \cap A]$

State any theorems used. (You may find Venn diagrams helpful.)

(i) *Proof:*

Assume that $A \subseteq B$. We show $C - B \subseteq C - A$:

Let $x \in C - B$. Then $x \in C$ and $x \notin B$.

Now $A \subseteq B \quad \therefore \quad x \notin B \Rightarrow x \notin A$.

$\therefore \quad x \in C$ and $x \notin A$.

$\Rightarrow x \in C - A$.

$\therefore \quad C - B \subseteq C - A. \square$

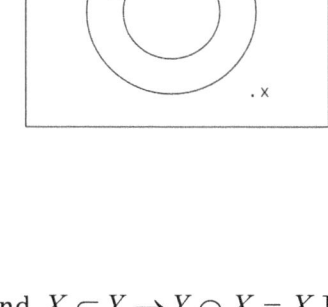

(ii) $[(A \cup B \cup C) \cap (A \cup B)] - [(A \cup (B - C)) \cap A]$

$= (A \cup B) - A \quad [A \cup B \subseteq A \cup B \cup C,\ A \subseteq A \cup (B - C)$ and $X \subseteq Y \rightarrow Y \cap X = X\,]$

$= (A \cup B) \cap A^c \quad$ [Theorem 2.4.1(x)]

$= (A \cap A^c) \cup (B \cap A^c) \quad$ [Intersection distributes over union on the right]

$= \emptyset \cup (B \cap A^c) \quad$ [Theorem 2.4.1(i)]

$= B - A \quad$ [Theorem 2.4.1(x)]. $\square$

Example 6: Prove using Set Algebra: $A - (A \cap B) = (A \cup B) - B$.

Proof:

$$A - (A \cap B) = A \cap (A \cap B)' \quad [\text{Theorem: } A - B = A \cap B']$$

$$= A \cap (A' \cup B') \quad [\text{De Morgan's Law}]$$

$$= (A \cap A') \cup (A \cap B') \quad [\text{Left Distributive Law}]$$

$$= \emptyset \cup (A \cap B') \quad [\text{Theorem: } A \cap A^c = \emptyset] \tag{1}$$

$$= A \cap B' \quad [\text{Theorem: } \emptyset \cup A = A].$$

$$(A \cup B) - B = (A \cup B) \cap B' \quad [\text{Theorem: } A - B = A \cap B']$$

$$= (A \cap B') \cup (B \cap B') \quad [\text{Right Distributive Law}]$$

$$= (A \cap B') \cup \emptyset \quad [\text{Theorem: } A \cap A^c = \emptyset]$$

$$= A \cap B' \quad [\text{Theorem: } A \cup \emptyset = A]$$

$$= A - (A \cap B) \quad [\text{from Equation (1)}]. \square$$

2.5 APPLICATION OF SET THEORY TO ARGUMENTS

You will be allowed to test the validity of arguments using Venn Diagrams.

Example 7:

Test the validity of the following argument:

P_1: All monkeys have tails

P_2: John does not have a tail

C: John is not a monkey

We can test the validity by recognising the argument or by using truth tables as we did in Chapter 1 (Exercise). We can also use Set Theory:

Let A be the set of monkeys and B be the set of those which have tails. Let j be John.

The given argument can be written:

P_1: $A \subseteq B$

P_2: $j \notin B$

C: $j \notin A$

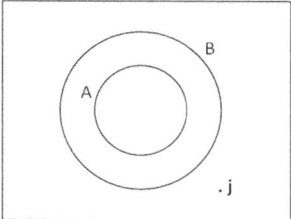

We can write the argument as a theorem in the language of sets and prove by contradiction:

Theorem 2.5.1: If $A \subseteq B$ and $j \notin B$ then $j \notin A$.

Proof:

Assume $j \in A$. Then $j \in B$ (since $A \subseteq B$). But $j \notin B$. This is a contradiction. $\therefore j \notin A$. $\square$

We can also use Venn diagrams to determine the validity of an argument. We must consider all possible diagrams where the premises are true. If there is one diagram where the conclusion is false then the argument is invalid.

Let us look first at how to illustrate some common statements with Venn diagrams.

Example 8:

Let S be the set of smart men, H be the set of handsome men and b be Brad.

In each case we translate the given statement into the language of sets and illustrate with Venn Diagrams:

(i). Brad is either smart or handsome: $b \in S \cup H$. There are 3 possible diagrams:

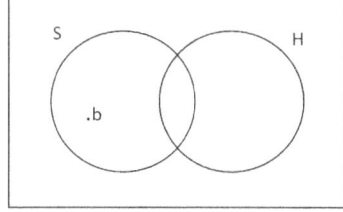 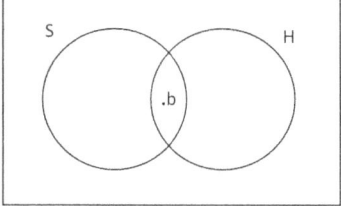 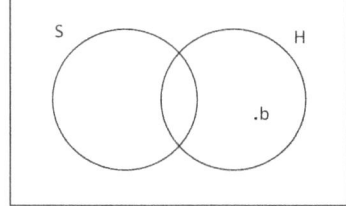

(ii) Brad is both smart and handsome: $b \in S \cap H$. (There is only one diagram.)

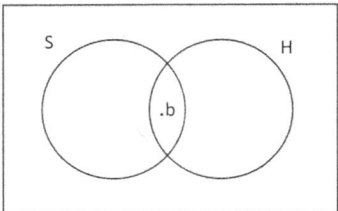

(iii). Brad is not handsome: $b \in H'$

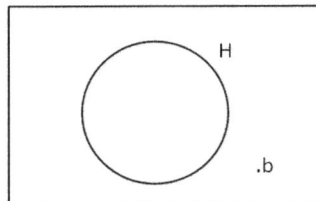

(iv). No handsome people are smart: $S \cap H = \emptyset$

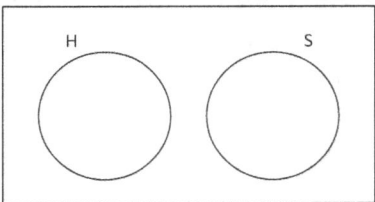

(v). Some handsome people are smart: $S \cap H \neq \emptyset$

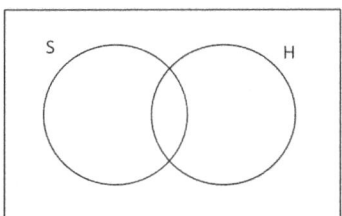

(vi). All handsome people are smart: $H \subseteq S$. (See diagram on next page).

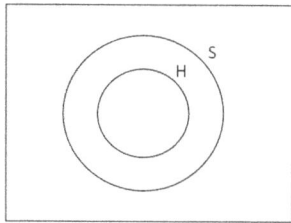

Exercise: Translate into the language of sets and draw Venn diagrams to represent the statement:

"If it is sunny, I will go to the beach."

Example 9: Use Venn diagrams to test the validity of the following argument:

P_1: Anyone who likes Mathematics is a nerd

P_2: Alex likes Mathematics

C: Alex is a nerd

Solution: Let M be the set of those who like Mathematics, N be the set of nerds and a represent Alex.

We can write the given argument in the language of sets:

P_1: $M \subseteq N$

P_2: $a \in M$

C: $a \in N$

There is only one possible diagram when the premises are true:

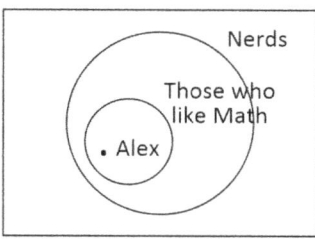

The diagram illustrates that whenever the premises are true, the conclusion is true. The argument is therefore valid.

Example 10: Test the validity of the following argument:

P_1: All students love holidays

P_2: Some who love holidays are lazy

C: Some students are lazy

Solution: Let A be the set of students, B be the set of holiday lovers and C be the set of lazy people.

In symbolic form:

P_1: $A \subseteq B$

P_2: $B \cap C \neq \emptyset$

C: $A \cap C \neq \emptyset$

We must consider all possible diagrams. If we meet up a case where the premises are true but the conclusion is false, then the argument is invalid.

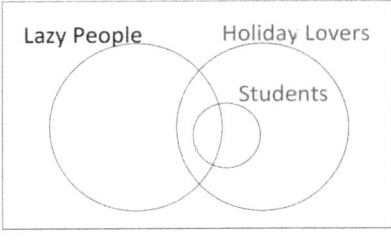
Diagram 1

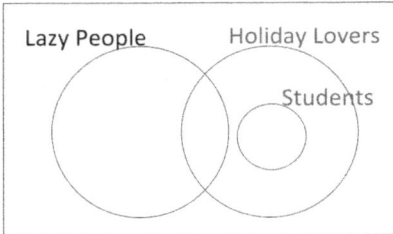
Diagram 2

Diagram 2 shows a case where the premises are true but the conclusion is false. The argument is therefore invalid.

2.6 FEATURE MATHEMATICIAN

JOHN VENN

"A PICTURE PAINTS A THOUSAND WORDS"

A painting of John Venn by C. E. Brock, 1899*

The idea of a set is basic and has been used in Mathematics since the days of Pythagoras (585-500 BC) but it was only within the last two hundred years that the theory was developed. This was primarily due to German mathematician, Georg Cantor (1870), who is considered the founder of Set Theory. The name more popularly associated with sets in introductory Mathematics courses is, however, John Venn. Though Swiss mathematician, Leonhard Euler (1768), had previously used diagrams to represent sets, it was Venn who developed and popularized these diagrams and earned the title to them.

John Venn (1834-1923) was descended from a long line of evangelical Christians. He obtained a degree in Mathematics from Gonville and Caius College, Cambridge, in 1857. He was ordained an Anglican priest two years later. In 1862, he returned to Cambridge, where he developed his diagrams. Venn wrote:

> I began at once somewhat more steady work on the subjects and books which I should have to lecture on. I now first hit upon the diagrammatical device of representing propositions by inclusive and exclusive circles. Of course the device was not new then, but it was so obviously representative of the way in which any one, who approached the subject from the mathematical side, would attempt to visualise propositions, that it was forced upon me almost at once.[1]

Venn developed Boole's and De Morgan's Mathematical Logic and made important contributions to the field. In 1903 he was elected president of Caius College and served there until his death in 1923. Although Venn left the priesthood to pursue his philosophical studies he remained, throughout his life, a man of sincere faith. A Venn diagram on the stained glass window of Caius Hall commemorates him.

* By kind permission of the Masters and Fellows of Gonville and Caius College, Cambridge.
[1]. Edwards, A. W. F. (2004). *Cogwheels of the Mind: The Story of Venn Diagrams.* JHU. p. 3.

2.7 CHAPTER EXERCISES

1a. Let A be a subset of B. Prove that

(i) $A \cap B = A$ (ii) $A \cup B = B$ (iii) $B' \subseteq A'$ (iv) $A \cup (B - A) = B$.

Illustrate with Venn Diagrams.

b. Let A, B and C be sets in a given universe. Prove formally:

(i) $(A - B) - C = (A - B) \cap (A - C)$

(ii) $A \cap (B \cup C) = (A \cap B) \cup (A \cap C)$

(iii) $A \cup (B \cap C) = (A \cup B) \cap (A \cup C)$

(iv) If $C \subseteq A$ then $A \cap (B \cup C) = (A \cap B) \cup C$, and conversely.

c. Prove formally that for all sets A and B in a given universe:

(i) $A \cup \emptyset = A$, $A \cap \emptyset = \emptyset$, $A \cap U = A$, $A \cap A^c = \emptyset$

(ii) $U^c = \emptyset$, $(A')' = A$

(iii) $A \subseteq A \cup B$, $A \cap B \subseteq A$

(iv) $A \cup B = B \cup A$.

2. Translate each of the following statements into the language of sets and draw appropriate Venn diagrams in each case:

a. Mathematics is both interesting and informative.

b. Men are the only animals which do not have tails.

3. Analyse the following argument using set theory. Begin by breaking the argument down into elementary arguments and express in the language of set theory:

> I do not have to see something to believe it. But if I see something I will believe it. I did not see anyone watering the plants in the yard, today. Therefore, I do not believe that the plants were watered.

4.(i) Let P and Q be non-empty sets. Prove that if P is a subset of Q, then the complement of Q is a subset of the complement of P.

(ii) Give an additional condition (not involving complements) which will be sufficient for the complement of Q to be equal to the complement of P.

(iii) Use your condition to formally prove that $P' = Q'$.

5. Use Set Algebra to obtain the following results for any sets S, T and W

(i) $T - (S \cap W) = (T - S) \cup (T - W)$

(ii) $(S - T) \cap T = \emptyset$.

6. (i) Give the general form of a Modus Tollens argument

(ii) Put the argument in set form

(iii) Use Venn diagrams to determine its validity.

7. Consider the following argument:

> If a man is a bachelor, he wears red shirts
> If a man wears black shirts, then he dies young
> People who wear black shirts do not wear red shirts
> ———————————————————————
> Bachelors die young

(i) Write the argument in symbolic form

(ii) Use Venn diagrams to determine its validity.

CHAPTER REFERENCES

E. J. Farrell, *M10A Lecture Notes* (Department of Mathematics and Computer Science, UWI, St. Augustine, 1998).

Michael C. Gemignani, *Basic Concepts of Math and Logic* (Mass: Addison-Wesley Publishing Co., Inc., 1968).

Seymour Lipschutz, *Schaum's Outline of Theory and Problems of Set Theory and Related Topics, Schaum's Outline Series* (New York : McGraw-Hill, 1977).

CHAPTER THREE

❖❖❖

BINARY OPERATIONS AND EQUIVALENCE RELATIONS

In this chapter we study some basic concepts in Mathematics. We first introduce the Cartesian product of two sets and then define a binary operation on a set. We study the commutative, associative, distributive and closure properties of binary operations, the existence of identity and inverse elements and also introduce some properties of real numbers. We then jump to equivalence relations—their reflexive, symmetric and transitive properties, equivalence classes and partitions—leaving functions for the next chapter. Technically, we should study relations and functions before studying binary operations but we have reversed the order, as many students find binary operations an easier topic and previous knowledge of functions is sufficient at that stage. One can, however, study chapter 4 before proceeding with this chapter.

3.1 BINARY OPERATIONS

Definition 3.1.1: Given two sets A and B, their *Cartesian product*, $A \times B$, is the set of all ordered pairs, whose first coordinate belongs to set A, and second, belongs to set B.

i.e. $A \times B = \{(x, y) : x \in A \text{ and } y \in B\}$.

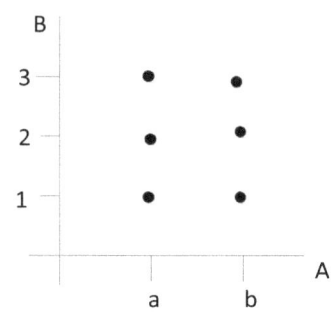

Example 1: $A = \{a, b\}$ $B = \{1, 2, 3\}$.

Then $A \times B = \{(a,1), (a,2), (a,3), (b,1), (b,2), (b,3)\}$.

Note that, in general, $A \times B \neq B \times A$.

Example 2: The Cartesian plane is the Cartesian product $\Re \times \Re$ where $\Re$ is the set of real numbers.

Definition 3.1.2: A *binary operation* $*$ on a set S is a rule which associates with each ordered pair (x, y) in $S \times S$ an element (denoted by $x * y$) which belongs to S.

Thus:
$$* : S \times S \to S$$
$$(x, y) \to x * y$$

The word 'binary' indicates that two elements are involved in the operation.

Example 3: The operation $+$ on the set $\Re$ of real numbers is defined by
$$+ : \Re \times \Re \to \Re$$
$$(x, y) \to x + y$$

For example, $+((2,3)) = 5$.

Definition 3.1.3: A binary operation $*$ on a set S is called *commutative* iff
$$x * y = y * x \text{ for all } x, y \in S.$$

Example 4: Consider the operation of intersection on a universal set S. Since $A \cap B = B \cap A, \forall A, B \in S$, $\cap$ is a commutative operation.

Example 5: The operation of subtraction on the set of integers Z is not commutative since, for example, $2 - 3 \neq 3 - 2$.

Definition 3.1.4: A binary operation $*$ on a set S is said to be *associative* iff
$$x * (y * z) = (x * y) * z \text{ for all } x, y, z \in S.$$

Example 6: Multiplication is associative on $\Re$ since $a \cdot (b \cdot c) = (a \cdot b) \cdot c$, $\forall x, y, z \in \Re$.

Definition 3.1.5: Let $*$ and Δ be two binary operations on a set A. The operation $*$ is said to *distribute* over the operation Δ iff, for every a, b, c in A,
$$a * (b \Delta c) = (a * b) \Delta (a * c).$$

Example 7: (i) $a \cdot (b + c) = (a \cdot b) + (a \cdot c)$, $\forall a, b, c \in \Re$. ($\cdot$ distributes over $+$).

(ii) $A \cap (B \cup C) = (A \cap B) \cup (A \cap C)$, i.e. $\cap$ distributes over $\cup$.

(iii) $a + (b \cdot c) \neq (a+b) \cdot (a+c)$, $\forall\, a,b,c \in \Re$. $+$ does not distribute over $\cdot$.

Example 8:

Let Δ be a binary operation on $\Re$, defined by $x \Delta y = x^2 + y^2$, $\forall\, x, y \in \Re$.

Determine whether or not the operation is (i) commutative (ii) associative.

Solution:

(i) Commutativity

We must show that $x \Delta y = y \Delta x$, $\forall\, x, y \in \Re$.

Well, $x \Delta y = x^2 + y^2$ and $y \Delta x = y^2 + x^2$.

But $x^2 + y^2 = y^2 + x^2$, $\forall\, x, y \in \Re$ (since $+$ is commutative).

$\therefore x \Delta y = y \Delta x$, $\forall\, x, y \in \Re$. $\therefore$ Δ is commutative.

(ii) Associativity

We must show that $(x \Delta y) \Delta z = x \Delta (y \Delta z)$, $\forall\, x, y, z \in \Re$.

$(x \Delta y) \Delta z = (x^2 + y^2) \Delta z = (x^2 + y^2)^2 + z^2 = x^4 + y^4 + 2x^2 y^2 + z^2$.

$x \Delta (y \Delta z) = x \Delta (y^2 + z^2) = x^2 + (y^2 + z^2)^2 = x^2 + y^4 + z^4 + 2 y^2 z^2$.

$(x \Delta y) \Delta z \neq x \Delta (y \Delta z)$, $\forall\, x, y, z \in \Re$. $\therefore$ Δ is not associative.

Definition 3.1.6: Let $*$ be defined on S. A subset B of S is *closed* under $*$ iff

$$x * y \in B, \ \forall\, x, y \in B.$$

Example 9: (i) $x + y \in \Re$, $\forall\, x, y \in \Re$. $\Re$ is closed under addition.

(ii) $x \cdot y \in \Re$, $\forall\, x, y \in \Re$. $\Re$ is closed under multiplication.

(iii) $x + y \in Z^+$, $\forall\, x, y \in Z^+$. The set of positive integers is closed under addition.

(iv) Z^+ is not closed under subtraction since, for example, $2, 3 \in Z^+$ but $2 - 3 = -1 \notin Z^+$. $\Re$ is, however, closed under subtraction.

Question: What about division? Is $\Re$ closed under $\div$? We must ask, does $x \div y \in \Re$, $\forall\, x, y \in \Re$? The answer is "No". For example, $2 \div 0 \notin \Re$, since it is undefined. $\Re$ is, therefore, not closed under $\div$ (N.B. a single counter example disproves a statement).

Note that as we have defined the term 'binary operation' on a set S, the set S is necessarily closed under the operation, since $x * y \in S$, $\forall\, x, y \in S$.

Example 10: Let Δ be an operation on $\Re$ defined by

$$x \Delta y = x^2 + y^2, \quad \forall\, x, y \in \Re.$$

Show that $\Re$ closed under Δ.

Solution:

Let $x, y \in \Re$. Then $x \Delta y = x^2 + y^2$.

Now $x \in \Re \rightarrow x^2 \in \Re$ (since $\Re$ is closed under multiplication).

Also $y \in \Re \rightarrow y^2 \in \Re$ (since $\Re$ is closed under multiplication).

But $x^2, y^2 \in \Re \rightarrow x^2 + y^2 \in \Re$ (since $\Re$ is closed under addition).

$\therefore x \Delta y \in \Re$, $\forall\, x, y \in \Re$. $\therefore \Re$ is closed under Δ.

Definition 3.1.7: Let $*$ be a binary operation on a set A. An element $e \in A$ is called an *identity* element for the operation $*$ iff, for every element $a \in A$,

$$e * a = a * e = a.$$

Definition 3.1.8: Let $*$ be a binary operation on a set A. Let $e \in A$ be the identity under $*$. Then the *inverse* of an element $a \in A$, denoted a^{-1}, is an element in A such that

$$a^{-1} * a = a * a^{-1} = e.$$

Example 11 (continued): $x \Delta y = x^2 + y^2$, $\forall\, x, y \in \Re$. We showed in Example 8, that Δ is commutative but not associative. Does $\Re$ have an identity element with respect to Δ?

Solution: e is identity iff $e \Delta x = x \Delta e = x$, $\forall x \in \Re$.

$e \Delta x = e^2 + x^2 = x$

$\Leftrightarrow e^2 = x - x^2 = x(1-x)$.

If $x > 1$ then e^2 is negative. This is impossible for $e \in \Re$.

$\therefore$ There is no identity element.

(Since there is no identity, we also cannot speak of inverse.)

Definition 3.1.9: Let $*$ be a binary operation on a set S, an element $e \in S$ is called a *right identity* for S iff $x * e = x$, $\forall x \in S$. e is called a *left identity* iff $e * x = x$, $\forall x \in S$.

If e is both a left and a right identity for S then e is an identity for S.

If an operation $*$ is commutative then left identity = right identity = identity.

Theorem 3.1.1: The identity, if it exists, is unique.

Proof:

Suppose that both e and f are identity elements. We show that $e = f$:

Since e is identity then $e * x = x * e = x$, $\forall x \in S$.

Since f is identity then $f * x = x * f = x$, $\forall x \in S$.

In particular since $f \in S$ and e is identity $e * f = f * e = f$.

Also since f is identity and $e \in S$, $f * e = e * f = e$.

Since $e * f = e$ and $e * f = f$, we have $f = e$ ($e * f$ is unique for a binary operation). $\square$

Definition 3.1.10: An element $b \in S$ is called a *right inverse* of $x \in S$ iff $x * b = e$. b is a *left inverse* iff $b * x = e$. An inverse is both right and left inverse.

Example 12: Consider the operation of multiplication on the set of real numbers.

$2 \cdot 1/2 = 1$ therefore $1/2$ is the right inverse of 2. $1/2 \cdot 2 = 1$, therefore $1/2$ is the left inverse of 2.

Theorem 3.1.2: Let $*$ be an associative operation, with an identity element. If b and b' are both inverses of an element a, then $b = b'$, (i.e. the inverse is unique if $*$ is associative).

Proof:

Let e be the identity element. $b = b * e = b * (a * b')$, since $b' = a^{-1}$.

Also, $b' = e * b' = (b * a) * b'$, since $b = a^{-1}$.

Now $b * (a * b') = (b * a) * b'$, if $*$ is associative. $\therefore$ $b = b'$, if $*$ is associative. $\square$

We have been using/assuming some basic properties of real numbers which have not been formally stated. In the next section, we list some properties of $\Re$, which you will be allowed to assume.

3.2 SOME PROPERTIES OF REAL NUMBERS

(i) $\Re$ is closed under $+$ and $\cdot$.

(ii) $(x+y)+z = x+(y+z)$ [$+$ is associative].

(iii) $(x \cdot y) \cdot z = x \cdot (y \cdot z)$ [$\cdot$ is associative].

(iv) $x \cdot (y+z) = x \cdot y + x \cdot z$ [left distributive]; $(x+y) \cdot z = x \cdot z + y \cdot z$ [right distributive].

(v) There exists a unique element $0 \in \Re$, such that
$$0 + x = x + 0 = x, \ \forall\, x \in \Re.$$

The element 0 is called the identity under $+$.

(vi) $\forall x \in \Re$, $\exists$ an element $-x \in \Re$, such that $x + -x = -x + x = 0$, the identity under $+$. $-x$ is called the additive inverse of x. Thus, every real number has an additive inverse.

(vii) There exists a unique element $1 \in \Re$:
$$1 \cdot x = x \cdot 1 = x, \ \forall x \in \Re.$$

1 is called the multiplicative identity of $\Re$.

(viii) For all non-zero real numbers, x, there exists a unique element $y \in \Re$, such that $x \cdot y = y \cdot x = 1$. The element $y = \dfrac{1}{x}$ is called the multiplicative inverse of x.

(For example,. $2 \cdot \frac{1}{2} = 1 = \frac{1}{2} \cdot 2$. Therefore $\frac{1}{2}$ is the multiplicative inverse of 2).

Question: Does every element of $\Re$ have a multiplicative inverse? Well, if $x = 0$, then $\dfrac{1}{x}$ is undefined. Therefore, every *non-zero* real number has a multiplication inverse.

Field Axioms

Let $(F,+,\cdot)$ be a non-empty set F; together with two (binary) operations *addition* (+) and multiplication ($\cdot$).

1. F is closed under + and $\cdot$
2. + is associative
3. $\cdot$ is associative
4. $\cdot$ distributes over +
5. $\exists$ an identity, 0, under +
6. Every element has an additive inverse
7. $\exists$ a multiplicative identity
8. Every $x \neq 0$ has a multiplicative inverse

Definition 3.2.1: Any non-empty set F, together with two operations $+, \cdot$ satisfying properties 1-8 is called a *field*.

The set of real numbers is therefore a field under $+, \cdot$. You may assume any of the eight properties in your proofs. There are other properties that are not included in this list. For example, + and $\cdot$ are commutative but these are not included as they can be derived from the others. Note that not all sets are fields. For example, Z is not a field under $+, \cdot$ since property 8 is not satisfied. You will learn more about fields if you take a more advanced course in Algebra.

3.3 WORKED EXAMPLES–BINARY OPERATIONS

Example 13:

$$a*b = 1 + 2ab - (a+b) \text{ on } \Re$$

Determine if (i) $\Re$ is closed under $*$ (ii) $*$ is commutative (iii) $*$ is associative (iv) an identity exists under $*$ (v) any elements have an inverse under $*$.

Solution: (i) Closure

$a, b \in \Re \rightarrow a + b \in \Re$ and $2ab \in \Re$ [$\Re$ is closed under + and $\cdot$]

$\rightarrow 1 + 2ab - (a+b) \in \Re$, $\forall a, b \in \Re$. Therefore $\Re$ is closed under $*$.

(ii) Commutativity

$a*b = 1 + 2ab - (a+b) = 1 + 2ba - (b+a)$, since + and $\cdot$ are commutative.

$\therefore a*b = b*a$, $\forall a, b \in \Re$. $\therefore *$ is commutative.

(iii) Associativity

$$(a*b)*c = [1+2ab-(a+b)]*c$$
$$= 1+2(1+2ab-(a+b))c-(1+2ab-(a+b)+c)$$
$$= 1+2c+4abc-2ac-2bc-1-2ab+a+b-c$$
$$= a+b+c-2ab-2ac-2bc+4abc$$

$$a*(b*c) = a*[1+2bc-(b+c)] = 1+2a(1+2bc-(b+c))-a-1-2bc+b+c$$
$$= a+b+c-2ab-2ac-2bc+4abc.$$

$(a*b)*c = a*(b*c)$, $\forall a,b,c \in \Re$. $\therefore$ * is associative.

(iv) Identity

e is the identity iff $a*e = e*a = a$, $\forall a \in \Re$.

$a*e = a$ iff $1+2ae-(a+e) = a$

iff $2ae - e = 2a - 1$

iff $e(2a-1) = 2a-1$.

$\therefore e = 1$ if $2a - 1 \neq 0$ or $a \neq 1/2$.

If $a = 1/2$, $1/2 * 1 = 1 + 2 \cdot 1/2 \cdot 1 - (1/2 + 1) = 1/2$.

$\therefore \forall a \in \Re$, $a*1 = a$.

Since * is commutative, $1*a = a$ also. $\therefore e = 1$ is the identity element.

(v) Inverse

An element y is the inverse of an element a, iff $a*y = y*a = e$.

Since * is commutative, we need only show $a*y = e$:

$a*y = 1 + 2ay - a - y = 1$
iff $y(2a-1) = a$
iff $y = a/(2a-1)$, $a \neq 1/2$

So, every element $a \in \Re$, except $a = 1/2$, has an inverse; namely $\dfrac{a}{2a-1}$. $\square$

It is sometimes convenient to define a binary operation by giving a table. Here is an example.

Example 14

Let $S = \{a,b,c,d\}$. Define an operation $\cdot$ on S, by the following table:

$\cdot$	a	b	c	d
a	a	b	c	d
b	b	c	d	a
c	c	d	a	b
d	d	a	b	c

(Note that the first row reads: $a \cdot a = a$, $a \cdot b = b$, $a \cdot c = c$, $a \cdot d = d$. Similarly for other rows).

(i) Closure: Clearly $x \cdot y \in S$, $\forall x, y \in S$. S is closed under $\cdot$.

(ii) The table is symmetric about the leading diagonal, therefore $x \cdot y = y \cdot x$, $\forall x, y \in S$. Therefore $\cdot$ is commutative.

(iii) Check that $(x \cdot y) \cdot z = x \cdot (y \cdot z)$ for all triples $x, y, z \in S$.

For example: $a \cdot (b \cdot c) = a \cdot d = d$ and $(a \cdot b) \cdot c = b \cdot c = d$.

You must check all triples in order to conclude that $\cdot$ is associative.

(iv) Row 1 and column 1 show that $a \cdot x = x = x \cdot a$, $\forall x \in S$. $\therefore a$ is the identity element.

(v) x is the inverse of a if $x \cdot a = a \cdot x = a$

$a \cdot a = a \rightarrow a^{-1} = a$
$b \cdot d = a \rightarrow b^{-1} = d$
$c \cdot c = a \rightarrow c^{-1} = c$
$d \cdot b = a \rightarrow d^{-1} = b$

Example 15: The binary operation $*$ is defined on the set $\Re$ of real numbers by
$$a * b = k + \frac{(a-k)(b-k)}{1-k}, \forall a, b \in \Re, \text{ where } k \neq 1 \text{ is a fixed real number.}$$

Show that (i) $*$ is commutative (ii) $*$ is associative (iii) $\Re$ contains an identity with respect to $*$.

Determine which, if any, elements have inverses.

Solution:

(i) Commutative

$$a*b = k + \frac{(a-k)(b-k)}{1-k} \quad \text{and} \quad b*a = k + \frac{(b-k)(a-k)}{1-k}.$$

But $k + \frac{(a-k)(b-k)}{1-k} = k + \frac{(b-k)(a-k)}{1-k}$, $\forall a, b \in \Re$ since $\cdot$ is commutative.

$\therefore$ $*$ is commutative.

(ii) Associative

$$(a*b)*c = [k + \frac{(a-k)(b-k)}{1-k}]*c$$

$$= k + \left(k + \frac{(a-k)(b-k)}{1-k} - k\right)\left(\frac{c-k}{1-k}\right)$$

$$= k + \frac{(a-k)(b-k)(c-k)}{(1-k)(1-k)}$$

$$a*(b*c) = a*[k + \frac{(b-k)(c-k)}{1-k}]$$

$$= k + \left(\frac{a-k}{1-k}\right)\left(k + \frac{(b-k)(c-k)}{1-k} - k\right)$$

$$= k + \frac{(a-k)(b-k)(c-k)}{(1-k)(1-k)}$$

$$= (a*b)*c, \; \forall a, b, c \in \Re$$

$\therefore$ $*$ is associative.

(iii) Identity

e is right identity if: $a*e = k + \frac{(a-k)(e-k)}{1-k} = a$, $\forall a \in \Re$.

Solving for e:

$$\therefore \frac{(a-k)(e-k)}{1-k} = a-k$$

$$\therefore (a-k)(e-k) = (a-k)(1-k)$$

$$\therefore e-k = 1-k, \quad a \neq k$$

$$\therefore e = 1.$$

When $a = k$, $k*1 = k + \frac{(k-k)(1-k)}{1-k} = k$

$\therefore a*1 = a, \forall a \in \Re$.

Since $*$ is commutative $a*e = e*a$. $\therefore 1*a = a, \forall a \in \Re$.

$\therefore e = 1$ is the identity element.

(iv) Inverse

x has inverse y iff $x*y = y*x = e$.

Since $*$ is commutative, we need only solve $x*y = e$, where $e = 1$.

$$x*y = k + \frac{(x-k)(y-k)}{1-k} = 1$$

$$\leftrightarrow \frac{(x-k)(y-k)}{1-k} = 1-k$$

$$\leftrightarrow (x-k)(y-k) = (1-k)^2$$

$$\leftrightarrow y - k = \frac{(1-k)^2}{x-k}, \quad x \neq k$$

$$\leftrightarrow y = k + \frac{(1-k)^2}{x-k}, \quad x \neq k.$$

If $x = k$, then y is undefined. But y must be an element of $\Re$. So every real number x, except $x = k$, has inverse; namely $k + \frac{(1-k)^2}{x-k}$.

Example 16: Let $S = \{x : 0 \leq x < 1\}$ and

$$x * y = \frac{x+y}{1+xy}, \quad \text{where } x, y \in S.$$

Determine the properties of $*$ and whether any elements of S have inverses under $*$.

(i) Closure

We must show that $x * y \in S, \ \forall \ x, y \in S$, i.e. $0 \leq x * y < 1, \ \forall x, y \in S$.

Clearly $x * y = \dfrac{x+y}{1+xy} \geq 0$, since $x + y \geq 0$ and $1 + xy > 0$, for $x, y \geq 0$.

We need only show that $x * y < 1$ or $1 - (x * y) > 0, \ \forall x, y \in S$:

$$1 - (x * y) = 1 - \frac{x+y}{1+xy} = \frac{1+xy-x-y}{1+xy} = \frac{1-x-y(1-x)}{1+xy} = \frac{(1-x)(1-y)}{1+xy} \quad (1)$$

But $0 \leq x < 1 \rightarrow 1 - x > 0$. Similarly $1 - y > 0$. Therefore the numerator of (1) is positive. Also $x, y \geq 0 \rightarrow xy \geq 0$. The denominator of (1) is thus positive. The fraction is therefore positive. $\therefore \ 1 - (x * y) > 0$ and $x * y < 1$.

Therefore S is closed under $*$.

(ii) Commutative

$$x * y = \frac{x+y}{1+xy} = \frac{y+x}{1+yx} = y * x, \ \forall x, y \in S. \text{ Therefore } * \text{ is commutative.}$$

(iii) Associative

$$x * (y * z) = x * \left(\frac{y+z}{1+yz}\right) = \frac{x + \dfrac{y+z}{1+yz}}{1 + x\left(\dfrac{y+z}{1+yz}\right)} = \frac{x + xyz + y + z}{1 + yz + xy + xz} = \frac{x + y + z + xyz}{1 + xy + xz + yz}.$$

Exercise: Verify that $(x * y) * z = \dfrac{x+y+z+xyz}{1+xy+xz+yz}$. $\therefore \ *$ is associative.

(iv) Identity

e is identity under $*$ iff $e*x = x*e = x$, $\forall x \in S$.

Since $*$ is commutative, it suffices that $e*x = x$, $\forall x \in S$.

$$e*x = \frac{e+x}{1+ex} = x \leftrightarrow e+x = x+ex^2 \leftrightarrow e(1-x^2) = 0.$$

Now, $x \neq 1$. $\therefore$ $1-x^2 \neq 0$. Therefore $e = 0$.

(iv) Inverse

y is the inverse of an element x, iff $x*y = y*x = e$.

$$\frac{x+y}{1+xy} = 0 \leftrightarrow x+y = 0 \leftrightarrow y = -x$$

$\therefore$ Only one element $x \in S$ has an inverse, namely $x = 0$; and $0^{-1} = (-0) = 0$.

3.4 EQUIVALENCE RELATIONS

Definition 3.4.1: A *relation* R on a set S, is a subset of $S \times S$.

A relation R is therefore a set of ordered pairs.

$(x, y) \in R$ iff xRy.

If $(x, y) \in R$, we write xRy (read: "x is related to y").

If x is not related to y, we write $(x, y) \notin R$.

Example 17:

Let $S = \{1, 2, 3, 4\}$. Define a relation R on S by

$$x\ R\ y \text{ iff } x < y.$$

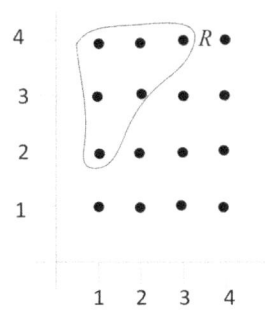

Then $R = \{(1, 2), (1,3), (1,4), (2,3), (2,4), (3,4)\}$.

Equivalence Relations

Definition 3.4.2: A relation R defined on a set S is called:

(i) *Reflexive* iff xRx, $\forall x \in S$.

(ii) *Symmetric* iff $xRy \to yRx$.

(iii) *Transitive* iff xRy and $yRz \to xRz$.

A relation which is reflexive on S, symmetric and transitive is called an *equivalence relation* on S.

Example 18 (Adapted from Enderton, *Elements of Set Theory*, p. 56.)

Consider a set A, for example, $A = \omega = \{0, 1, 2, 3, ...\}$. We want to partition A into 6 little boxes:

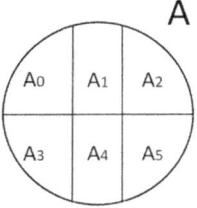

One such partition is : $A_0 = \{0, 6, 12, ...\}$, $A_1 = \{1, 7, 13, ...\}$, ... $A_5 = \{5, 11, 17, ...\}$.

By a partition we mean that every element of A is in exactly one little box and that each box is a non-empty subset of A. Now we can think of each little box as a single object. The set: $B = \{A_0, A_1, ..., A_5\}$ has only six members; whereas A is infinite. It is thus easier to handle.

Now define a relation R on A as follows:

For $x, y \in A$, $x R y \Leftrightarrow x$ and y are in the same little box.

For example, 0 is related to 6 but 0 is not related to 1. We can see that R has the following properties:

1. $x R x$ for all $x \in A$, i.e. x and x are in the same little box.

2. Whenever $x R y$ then $y R x$. For example, $6R12 \leftrightarrow 12R6$.

3. Whenever $x R y$ and $y R z$ then $x R z$. For example, $0R6$ and $6R12 \to 0R12$.

As R satisfies these 3 properties it is an equivalence relation. We say that the partition on the set A, induces the equivalence relation.

Example 19:

Let R be the relation '<' on the set $\Re$.

(i) Is $x < x$? No! $\therefore$ R is not reflexive. (ii) $x < y \nrightarrow y < x$. $\therefore$ R is not symmetric.

(iii) $x < y$, $y < z \rightarrow x < z$. $\therefore$ R is transitive. Hence '<' is not an equivalence relation on $\Re$.

Example 20:

Let R be the relation '$\leq$' on the set $\Re$.

(i) $x \leq x$, $\forall x \in \Re$. $\therefore$ R is reflexive. (ii) $x \leq y \nrightarrow y \leq x$. $\therefore$ R is not symmetric.

(iii) $x \leq y$ and $y \leq z \rightarrow x \leq z$. $\therefore$ R is transitive. Hence '$\leq$' is not an equivalence relation on $\Re$.

Example 21:

Now consider the relation of 'equality' on the set $\Re$:

$$xRy \text{ iff } x = y.$$

(i) $x = x$, $\forall x \in \Re$. $\therefore$ R is reflexive. (ii) $x = y \rightarrow y = x$. $\therefore$ R is symmetric.

(iii) $x = y$ and $y = z \rightarrow x = z$. $\therefore$ R is transitive. Hence '=' is an equivalence relation on $\Re$.

Equality is considered to be the simplest of all equivalence relations.

Example 22:

A relation R is defined on the set Z of integers as follows

$$\forall \ a, b \in Z, \ aRb \text{ iff } a - b \text{ is an even integer.}$$

Show that R is an equivalence relation on Z.

Proof:

(i) To show that R is reflexive we show that aRa, $\forall a \in Z$.

$a - a = 0$, which is an even integer. $\therefore$ aRa, $\forall a \in Z$ $\therefore$ R is reflexive.

(ii) Symmetric: We show that $aRb \rightarrow bRa$:

Assume that aRb for some $a,b \in Z$. Then, $a-b$ is an even integer. Now $b-a = -(a-b)$. ∴ $b-a$ is also an even integer. ∴ $aRb \rightarrow bRa$. ∴ R is symmetric.

(iii) Transitive: We show that $a-b$ and $b-c$ even integers imply that $a-c$ is an even integer:

Let aRb and bRc for some $a,b,c \in Z$. Then $a-b$ and $b-c$ is are even integers.

∴ $a-b = 2z_1$, $z_1 \in Z$ and $b-c = 2z_2$, $z_2 \in Z$.

∴ $a-b+b-c = 2(z_1 + z_2) = 2z_3$ where $z_3 \in Z$.

∴ aRc and R is transitive

Since R is reflexive, symmetric and transitive, it is an equivalence relation on Z. □

Example 23: Let $S = \{a,b,c\}$. Then

$R_1 = \{(a,a),(b,b),(c,c),(a,b),(b,a),(c,a),(a,c),(b,c),(c,b)\}$

is an equivalence relation on S. However,

$R_2 = \{(a,a),(b,b),(c,c),(a,b),(b,a),(c,a),(a,c),(b,c)\}$ is not;

since $(b,c) \in R_2$, but $(c,b) \notin R_2$ (so R_2 is not symmetric).

$R_3 = \{(a,a)(b,b)(c,c)\}$ is also an equivalence relation but $R_4 = \{(a,a)(b,b)\}$ is not (since R_4 is not reflexive).

Exercise: List all the equivalence relations on the set $\{a,b,c\}$.

We can find many examples of equivalence relations.

Example 24:

The relation of congruence of triangles is an equivalence relation. Recall that congruent triangles have the same size and shape.

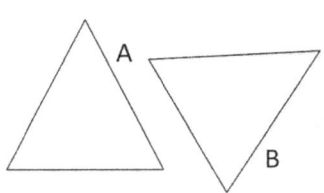

A is congruent to B

Let $S = \{\text{all triangles}\}$

Every triangle is congruent to itself. ∴ $A \equiv A, \forall A \in S$.

If $A \equiv B$ then $B \equiv A$. If $A \equiv B$ and $B \equiv C$ then $A \equiv C$.

Example 25

The relation 'is similar to' is an equivalence relation on the set of triangles. Recall that A is similar to B if their sides are in a corresponding ratio.

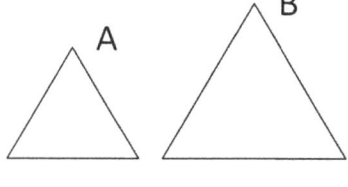

A is similar to B

Exercise: Prove that the relation 'is similar to' is an equivalence relation on the set of triangles.

Example 26:

Recall that two vectors are equal iff they have the same length and direction. Equality of vectors is an equivalence relation.

Even if vectors are not equal in length; but just parallel, we have an equivalence relation on the set of lines in a plane P.

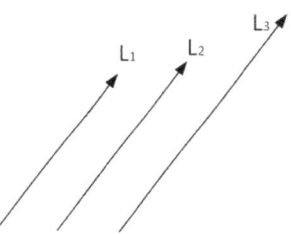

$L \| L \ \forall \ L \in P$.

$L_1 \| L_2 \rightarrow L_2 \| L_1$.

$L_1 \| L_2$ and $L_2 \| L_3 \rightarrow L_1 \| L_3$.

Parallel lines

Example 27:

Let $W = \{0, 1, 2, ...\}$. A relation $\sim$ is defined on W by

$$m \sim n \Leftrightarrow m - n \text{ is divisible by } 6.$$

Show that $\sim$ is an equivalence relation (and find the set of equivalence classes).

Solution:

(i) $\forall \, m \in W, m \sim m$, since $m - m = 0$, which is divisible by 6. $\therefore \sim$ is reflexive.

(ii) $m \sim n \rightarrow m - n$ is divisible by 6

$\Rightarrow n - m = -(m - n)$ is divisible by 6

$\Rightarrow n \sim m . \therefore \sim$ is symmetric.

(iii) $m \sim n$ and $n \sim p \Rightarrow m - n$ is divisible by 6 and $n - p$ is divisible by 6

$\Rightarrow m - n = 6 z_1, \ z_1 \in Z$ and $n - p = 6 z_2, \ z_2 \in Z$.

$\Rightarrow m - n + n - p = m - p = 6(z_1 + z_2) = 6z_3$ where $z_3 \in Z$.

$\Rightarrow m \sim p$. $\therefore \sim$ is transitive.

Before we can finish the example we must study equivalence classes.

3.5 EQUIVALENCE CLASSES

Definition 3.5.1: Let R be an equivalence relation on a set S, then the *equivalence class* of an element x, written $[x]$, is $\{y \in S : xRy\}$, i.e. the set of all elements which are related to x.

Note that since R is an equivalence relation, and $xRy \rightarrow yRx$, we may also say

$$[x] = \{y \in S : yRx\}.$$

Example 27 (continued): Consider further the equivalence relation in Example 27. Define the following:

$$A_0 = \{m \in W : (m-0) \text{ is divisible by } 6\}$$
$$A_1 = \{m \in W : (m-1) \text{ is divisible by } 6\}$$
$$A_2 = \{m \in W : (m-2) \text{ is divisible by } 6\}$$
$$A_3 = \{m \in W : (m-3) \text{ is divisible by } 6\}...$$
$$A_x = \{m \in W : mRx\}$$

Then:

$A_0 = \{0, 6, 12,...\}$ $A_6 = \{6, 12, 0,...\} = A_0 = A_{12}...$
$A_1 = \{1, 7, 13,...\}$ $A_7 = \{1, 7, 13,...\} = A_1 = A_{13}...$
$A_2 = \{2, 8, 14,...\}$ $A_8 = A_2...$
$A_3 = \{3, 9, 15,...\}$ $A_9 = A_3...$
$A_4 = \{4, 10, 16,...\}$ $A_{10} = A_4...$
$A_5 = \{5, 11, 17,...\}$ $A_{11} = A_5...$

There are six distinct sets: $A_0, A_1, ..., A_5$. Each element of W will be in one and only one of these six sets. A_0 is the equivalence class of 0. It is the same as the equivalence class of 6, etc. Therefore, $[0] = \{0, 6, 12,...\} = A_0$, $[1] = \{1, 7,...\} = A_1$, etc.

Example 28:

R is the relation 'equality' on $\Re$. The equivalence class of $x = \{y \in \Re : x = y\} = \{x\}$. So $[1] = \{1\}$, $[2] = \{2\}$, etc. Each element is its own equivalence class. (See Example 21).

Example 29:

In Example 22, we looked at the equivalence relation R on Z (the set of integers); defined by:

$$aRb \text{ iff } a - b \text{ is even.}$$

Then, $[0] = \{y \in Z : 0Ry\} = \{y : 0 - y \text{ is even}\} = \{...,-4, -2, 0, 2, 4,...\}$.

$\therefore [0]$ is the set of even integers.

$[1] = \{y \in Z : y - 1 \text{ is even}\} = \{...,-3, -1, 1, 3,...\}$.

$\therefore [1]$ is the set of odd integers.

We have, $[2] = [0] = [4] = [-2] = [-4]...$ and $[3] = [1] = [5] = [-1] = [-3]...$

Notice that each element is a member of its equivalence class and that if two elements are related then their equivalence classes are equal. There are only two distinct equivalence classes. They break up the set into two disjoint parts, the union of which gives Z.

Definition 3.5.2: A *partition* of X is a disjoint collection of non-empty subsets of X whose union is X.

Example 30: $X = \{a,b,c,d,e,f,g\}$. Let $C_1 = \{a,b,e,g\}$, $C_2 = \{c\}$, and $C_3 = \{d,f\}$. $\rho = \{C_1, C_2, C_3\}$ is a partition of X. But if $A_1 = \{a,c,e\}$, $A_2 = \{b,c\}$, and $A_3 = \{d,g\}$, then $\{A_1, A_2, A_3\}$ is not a partition since the element f is left out and c is in two sets.

An equivalence relation induces a partition on the set, namely, the set of equivalence classes. If R is an equivalence relation on a set X, then the set of equivalence classes is a partition of X.
In Example 29, $\{[0], [1]\}$ is a partition of Z.

The equivalence class of $x \in X$, i.e. $[x]_R$, is also denoted x/R. Also, we write X/R (read 'X modulo R') to denote the set of all equivalence classes of a set X, with respect to R.

In Example 29, $Z/R = \{[0], [1]\}$. In Example 27, $W/\sim = \{[0], [1], ..., [5]\}$.

We have already noticed in Example 29 that and if two elements were related then their equivalence classes were equal, e.g. $0R2$ and $[2]=[0]$. We now formally state this result.

Theorem 3.5.1: If R is an equivalence relation on a set A and $x, y \in A$ then $[x]=[y]$ iff xRy.

Proof: We must show two things:

(i) $[x]=[y] \to xRy$ and (ii) $xRy \to [x]=[y]$.

Proof of (i): Assume $[x]=[y]$.

We know xRx, $\therefore$ $x \in [x]$. Also, $y \in [y] = [x]$, by assumption. $\therefore$ $y \in [x]$.

$\therefore xRy$, by definition of equivalence class.

Proof of (ii): Now assume xRy.

To show that $[x]=[y]$ we show $[x] \subseteq [y]$ and $[y] \subseteq [x]$.

Well, $t \in [y] \Rightarrow yRt$. So we have xRy and yRt. $\therefore xRt$, since R is an equivalence relation.

$\therefore t \in [x]$. $\therefore [y] \subseteq [x]$.

Now, let $t \in [x]$. Then xRt. Also, yRx [since xRy by assumption]. But yRx, $xRt \Rightarrow yRt$. $\therefore t \in [y]$. $\therefore [x] \subseteq [y]$.

Since $[x] \subseteq [y]$ and $[y] \subseteq [x]$, we have $[x]=[y]$. □

We saw that an equivalence relation induces a partition on the set. Conversely, a partition on a set also induces an equivalence relation.

Example 31

$A = \{a,b,c\}$ and $\rho = \{\{a,b\},\{c\}\}$ – a partition of A. Find an equivalence relation R which induces ρ.

Solution: Define R to be such that xRy iff x, y are in same set. Then aRb, bRa, aRa, bRb and cRc. $R = \{(a,a),(b,b),(c,c),(a,b),(b,a)\}$ is the equivalence relation which induces the partition, ρ.

(NB. A partition on a set is not unique. In the chapter exercises you will be asked to write all the partitions of the set A and their corresponding equivalence relations.)

3.6 WORKED EXAMPLES–EQUIVALENCE RELATIONS

Question 1. Let Z be the set of integers and $S = Z \times Z$. Define a relation ρ on S by

$$(a,b) \rho (c,d) \text{ iff } a+d=b+c.$$

Show that ρ is an equivalence relation, and determine the equivalence class of $(2,5)$.

Solution:

(i) ρ is reflexive iff $(a,b) \rho (a,b)$, $\forall (a,b) \in Z \times Z$.

Well, $a+b = b+a, \forall a,b \in Z$. $\therefore (a,b) \rho (a,b)$ and therefore ρ is reflexive.

(ii) To show $(a,b) \rho (c,d) \rightarrow (c,d) \rho (a,b)$:

Let $(a,b) \rho (c,d)$. Then $a+d = b+c \rightarrow c+b = d+a \rightarrow (c,d) \rho (a,b)$. Therefore ρ is symmetric.

(iii) Let $(a,b) \rho (c,d)$ and $(c,d) \rho (e,f)$. Then $a+d = b+c$ and $c+f = d+e$

$\Rightarrow a+d+c+f = b+c+d+e$

$\Rightarrow a+f = b+e$. $\therefore (a,b) \rho (e,f)$. $\therefore \rho$ is transitive. Hence ρ is an equivalence relation.

$[(2,5)] = \{(a,b) \in Z \times Z : (2,5) \rho (a,b)\} = \{(a,b) \in Z \times Z : 2+b = 5+a\}$

$= \{(a,b) \in Z \times Z : b-a = 3\} = \{...,(-2,1),(-1,2),(0,3),(1,4),(2,5),...\}$.

Note, we had to consider the negative integers also, since $S = Z \times Z$. If $S = Z^+ \times Z^+$, then

$[(2,5)] = \{(x,y) \in Z^+ \times Z^+ : (x,y) R (2,5)\} = \{(x,y) \in Z^+ \times Z^+ : y - x = 3\} = \{(1,4),(2,5),...\}$.

Question 2. Let $\Re^+$ be the set of positive real numbers. Define a relation R on $\Re^+$ as follows:

$$xRy \Leftrightarrow \log_{10} x - \log_{10} y = k, \text{ where } k \in Z.$$

Show that R is an equivalence relation on $\Re^+$ and determine $[x]$.

Solution:

(i) $\forall x \in R^+$, $\log_{10} x - \log_{10} x = 0 = k \in Z$. $\therefore xRx$. $\therefore R$ is reflexive.

(ii) Suppose xRy. Then $\log_{10} x - \log_{10} y = k$, for some $k \in Z$.

$\Rightarrow \log_{10} y - \log_{10} x = -k$, where $-k \in Z$. $\therefore yRx$. $\therefore R$ is symmetric.

(iii) Suppose xRy and yRz for some $x, y, z \in \Re^+$. Then

$\log_{10} x - \log_{10} y = k_1$ and $\log_{10} y - \log_{10} z = k_2$, where $k_1, k_2 \in Z$

$\Rightarrow \log_{10} x - \log_{10} z = k_1 + k_2 = k_3 \in Z$. $\therefore xRz$. $\therefore R$ is transitive.

Since R is reflexive, symmetric and transitive, it is an equivalence relation.

$[x] = \{y \in \Re^+ : xRy\} = \{y \in \Re^+ : \log_{10} x - \log_{10} y = k \in Z\}$.

$= \left\{y \in \Re^+ : \log_{10} \dfrac{x}{y} = k\right\}$.

But $10^k = \dfrac{x}{y} \Rightarrow y = x10^k$.

$\therefore [x] = \{y \in \Re^+ : y = x10^k, k \in Z\} = \{\ldots, 10^{-2}x, 10^{-1}x, x, 10x, 10^2 x, \ldots\}$.

Question 3. Let $\aleph = \{0, 1, 2, \ldots\}$. Let $xRy \Leftrightarrow 3|(x-y)$, $x, y \in \aleph$. Show that R is an equivalence relation and find the set of equivalence classes.

Solution:

(i) Let $x \in \aleph$. Then $x - x = 0 = 3 \cdot 0$. $\therefore xRx, \forall x \in \aleph$. $\therefore R$ is reflexive.

(ii) $xRy \to 3|(x-y) \to x - y = 3k$, where $k \in Z$.

$\to y - x = -3k = 3k_1$, where $k_1 = (-k) \in Z$.

$\to yRx$. $\therefore R$ is symmetric.

(iii) Suppose that xRy and yRz. Then $x - y = 3k_1$ and $y - z = 3k_2$, for some $k_1, k_2 \in Z$.

Adding these equations, we get $x - z = 3(k_1 + k_2) = 3k_3$, $k_3 \in Z$. $\therefore xRz$. $\therefore R$ is transitive.

Since R is reflexive, symmetric and transitive it is equivalence.

The set of equivalence classes $\aleph/R = \{[0], [1], [2]\}$, where $[0] = \{0, 3, 6, \ldots\}$, $[1] = \{1, 4, 7, \ldots\}$ and $[2] = \{2, 5, 8, \ldots\}$.

3.7 FEATURE MATHEMATICIAN

RENÉ DESCARTES

"FATHER OF CARTESIAN GEOMETRY"

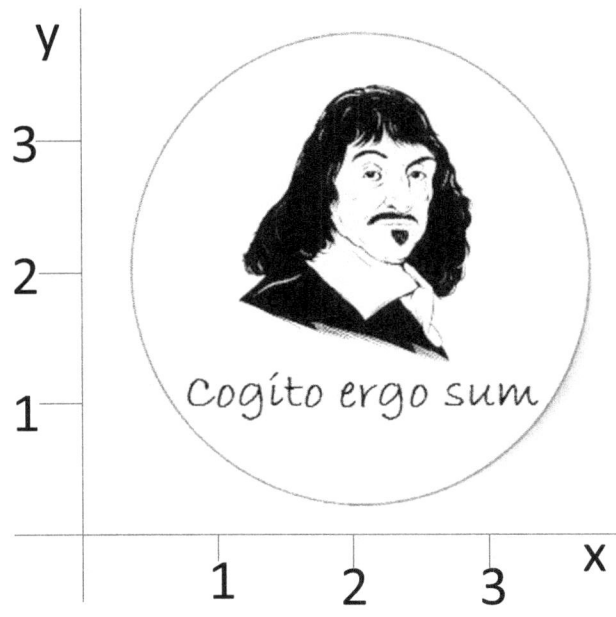

René Descartes (1596-1650) was a French philosopher, mathematician, and scientist. He is credited as the father of analytical geometry. The Cartesian coordinate system (Cartesian Product, $\Re \times \Re$)— allowing reference to a point in space as a set of numbers, and allowing algebraic equations to be expressed as geometric shapes in a two- or three-dimensional coordinate system (and vice versa)— was named after him. Descartes' ground-breaking work—the bridge between geometry and algebra, allowed the development of Newton's and Leibniz's infinitesimal calculus.

History has it that one night in November 1619, Descartes shut himself up, in a room with an "oven" to escape the cold. While within the room, he experienced visions and upon exiting formulated analytical geometry and the idea of applying the mathematical method to philosophy. Descartes saw that all truths were linked with one another, so that finding a fundamental truth and proceeding with logic, would open the way to all science. For Descartes, this basic truth is encapsulated in his now hallmark statement, "I think, therefore I am."[1]

Descartes used this truth to formulate a proof for the existence of God. He wrote, "The idea of God, or a supremely perfect being, is one that I find within me just as surely as the idea of any shape or number."[2] Existence, then, is necessarily derived from this clear and distinct idea. Descartes' philosophical reasoning for the existence of God laid the foundation for 17th century rationalism and subsequent modernity. However, he could not have realized the extent of his revolutionary gesture, as he considered himself a devout Catholic; and one of the main purposes of his *Meditations* was to defend the Christian faith.

Descartes made many other contributions to science; and was undoubtedly a genius of the first magnitude. For 2000 years Euclidean geometry lay at a standstill, until he developed his analytic geometry. He most likely had more influence than anyone else on the young Newton, who went on, not long after Descartes' death, to discover calculus and the entire framework of modern science.

1. "René Descartes," *Wikipedia*, https://en.wikipedia.org/wiki/Ren%C3%A9_Descartes.

2. Nolan, Lawrence, "Descartes' Ontological Argument", *The Stanford Encyclopedia of Philosophy*, http://plato.stanford.edu/entries/descartes-ontological/.

3.8 CHAPTER EXERCISES

1. Let $S = \{a,b,c,d\}$. Define an operation $*$ on S, by the following table:

*	a	b	c	d
a	d	a	c	b
b	a	c	b	d
c	b	d	a	c
d	c	b	d	a

Determine the algebraic properties of $*$. (Remember that a counter-example disproves a statement).

2. Let $P(X)$ be the set of all subsets of a set X. Let $*$ be the operation of intersection on $P(X)$, that is,
$$\forall A, B \in \rho(X), \quad A * B = A \cap B .$$
(i) Determine the algebraic properties of $*$.
(ii) Which, if any, elements of X have inverses under $*$?

3. Let $*$ be an operation defined on Z (the set of integers) by
$$a * b = a + b - ab, \text{ for all } a, b \in Z .$$
(a) Determine the following:
   (i). Is Z closed under $*$?
   (ii). Is $*$ commutative?
   (iii). Is $*$ associative?
(b) Does Z have an identity element with respect to $*$?
(c) Do any elements of Z have inverses under $*$?

4. Let S be the set of fractions, excluding -1. A binary operation $*$ is defined on S by
$$x * y = x + y + xy, \text{ for all } x, y \in S.$$
(a). Determine whether:
   (i) $*$ is commutative
   (ii) $*$ is associative
   (iii) S has an identity with respect to $*$.
(b). Determine which elements of S have inverses with respect to $*$.

5. Let $X = \Re \times \Re$, where $\Re$ is the set of real numbers. Let @ be a binary operation on the set X defined by
$$(a,b)@(c,d) = (ac, ad+bc) \text{ for all } (a,b) \text{ and } (c,d) \text{ in } X.$$
(i) Show that @ is commutative.
(ii) Determine whether or not @ is associative.
(iii) Show that X has an identity element under @.
(iv) Determine which elements (if any) of X have inverses under @; and find the inverse of each such element.

6. (a) Let $S = \{a,b,c\}$. Determine which of the following are equivalence relations on S:
   (i) $R_1 = \{(a,a),(b,b),(c,c)\}$
   (ii) $R_2 = \{(a,a),(b,b),(c,c),(a,b),(b,a)\}$
(b). List all the partitions of the set S, as well as, their corresponding equivalence relations.

7. Let n be a fixed positive integer. Define a relation P on Z (the set of integers) by
$$a \, P \, b \quad \text{iff} \quad b - a = np, \text{ where } p \in Z.$$
(i) Prove that P is an equivalence relation.
(ii) Find the equivalence classes of [0], [1] and [a], where $a \in Z$. Find, also, the set of equivalence classes.

8. Define the following terms with reference to a relation R defined on a set A:
   (i) Reflexive on A
   (ii) Symmetric
   (iii) Transitive

A relation R on a set A is called "circular" if aRb and bRc imply cRa, for all a, b, c in A. Let R be a relation on the set Z of integers defined by:
$$aRb \quad \text{iff} \quad a+3-b \text{ is divisible by } 9.$$
Show that R is a circular relation but that it is not an equivalence relation.

9. Let S be the set of ordered pairs of integers (a, b), with $b \neq 0$. A relation R is defined on S by
$$(a,b)R(c,d) \text{ iff } ad = bc.$$
(i) Prove that R is an equivalence relation on S.
(ii) Determine the equivalence class of $(7, 3)$.

10. Let R be a relation on $\mathbb{Z}$ (the set of integers) defined by
$$xRy \Leftrightarrow 3|(x-y), \text{ (i.e. 3 divides } (x-y)).$$
Show that R is an equivalence relation and find all the equivalence classes.

11. A relation is said to be *left Euclidean* iff
$$xRz \text{ and } yRz \rightarrow xRy.$$
Prove that if a relation is left Euclidean and reflexive, then it is an equivalence relation.

CHAPTER REFERENCES

E. J. Farrell, *M10A Lecture Notes* (Department of Mathematics and Computer Science, UWI, St. Augustine, 1998).

Herbert B. Enderton, *Elements of Set Theory* (New York: Academic Press, 1977).

Seymour Lipschutz, *Schaum's Outline of Theory and Problems of Set Theory and Related Topics*, Schaum's Outline Series (New York : McGraw-Hill, 1977).

CHAPTER FOUR

FUNCTIONS

In this chapter we learn about one of the most fundamental concepts in Mathematics—the notion of a function. We investigate the injective and surjective properties of functions, graphical tests for these properties and their formal proofs. We then move on to composition of functions, the existence of inverse functions and the computation of the inverse; when it exists.

4.1 DEFINITION OF A FUNCTION

Recall that a relation is defined as a set of ordered pairs. A function is a special type of a relation. For example, Let $A = \{\text{Tom, Bob, Rob}\}$ and $B = \{\text{Thomas, Robert, Joseph}\}$. Consider the relation $R = \{(x, y) : y \text{ is the father of } x\}$ given by:

$$R = \{(\text{Tom, Thomas}), (\text{Bob, Robert}), (\text{Rob, Robert})\}.$$

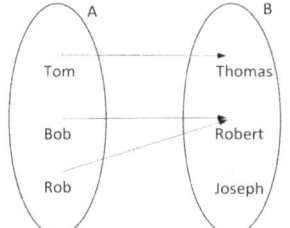

Notice that each man has only one father, two men may have the same father and not every man has a son. The relation 'is the father of' is a function. Not all relations are functions, however. Here is the definition.

Definition 4.1.1: Let A and B be sets. A *function* (mapping) from A to B is a relation which assigns to each element in a set A, a unique element of a set B.

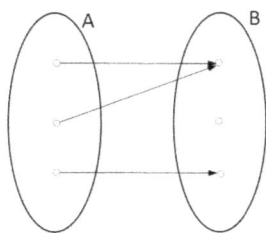

A Typical Function

Recall that a binary operation $*$ on a set A is a rule which associates with each ordered pair (x, y) in $A \times A$ an element $x * y$ which belongs to A. A binary operation is thus a function from $A \times A$ into A. Technically, we should study functions before we study binary operations but we have reversed the order, as most students find binary operations an easier topic and your previous knowledge of functions was sufficient at that stage. You should, however, review the material on binary operations, in light of the theory we now develop on functions.

Typically, we let f denote the function assignment and write $f : A \to B$; which reads 'f is a function from A into B'.

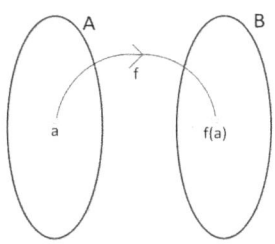

The set A is called the *domain* of the function (denoted $\text{Dom } f$) and B is called the *co-domain* of f.

For each object $a \in A$, the element in B to which it is assigned is called the *image* of a under f, denoted $f(a)$, and read, 'f of a'. This function notation was first introduced by Swiss Mathematician, Leonhard Euler. You can read about him at the end of the chapter.

Note that there may be some elements in B which are not the image of any element in A; but each element in A must have an image in B. A function must also be single valued. The following diagrams do *not* illustrate functions:

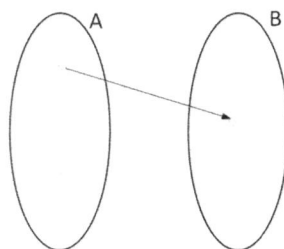

Every element in A must have an image

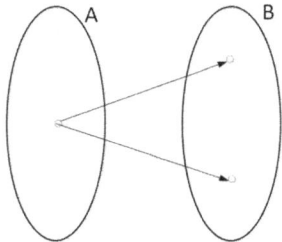

One to many is not a function

FUNCTIONS

Definition 4.1.2: The subset of elements in B which are the images of elements in A, is called the *range* of f. It is denoted $f(A)$ or *Ran f*.

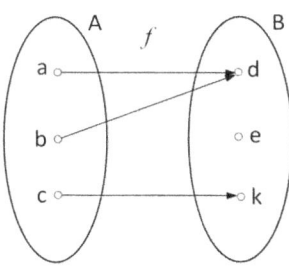

Example 1: $A = \{a,b,c\}$ and $B = \{d,e,k\}$.

$f = \{(a,d),(b,d),(c,k)\}$.

$Ran\ f = \{d,k\}$

[Note that $f(A) \subseteq B$].

Example 2:

$f : \Re \to \Re$ such that $x \to x^2$ is a function. (Every real number has a unique square). $Ran\ f = \Re^+ \cup \{0\}$.

x	-2	-1	0	1	2	3
x²	4	1	0	1	4	9

Example 3:

$f : \Re \to \Re$ such that $x \to \sqrt{x}$ is not a function, since a negative number has no real square root and a positive number has two square roots.

x	-4	-1	0	1	4
√x	?	?	0	±1	±2

For the graph a function every *vertical line* must cut the graph once and only once.

$\pm\sqrt{x}$ is not a function

To define a square root function, we would have to restrict the domain and codomain.

Definition 4.1.3: Let $f: A \to B$ and $g: C \to D$. If $A \subseteq C$, $B \subseteq D$ and $f(x) = g(x)$, $\forall x \in A$, then the function f is called the *restriction* of g to A, written g/A, while g is called the *extension* of f to C.

Example 4:

Let $f: \Re^+ \to \Re$ be such that $x \to x^2$ and $g: \Re \to \Re$ be such that $x \to x^2$. Then f is the restriction of g to $\Re^+$ and g is the extension of f to $\Re$.

Definition 4.1.4: Let $f: A \to B$ and $g: A \to B$. Then f and g are *equal* iff $f(a) = g(a)$, for every $a \in A$.

Example 5:

Let $f: \Re \to \Re$ be such that $f(x) = x^2$ and $g: \Re \to \Re$ be such that $g(y) = y^2$. Then f and g are equal (x and y are "dummy" variables).

Recall that the range of a function is always a subset of the co-domain. When the range is equal to the codomain, we have the special case presented next.

4.2 ONTO FUNCTIONS

Definition 4.2.1: Let f be a function from A into B. If every element of B is the image of some element of A, we say that f is a function from A onto B (i.e. $\text{Ran } f = B$).

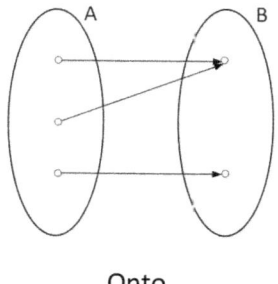

Onto

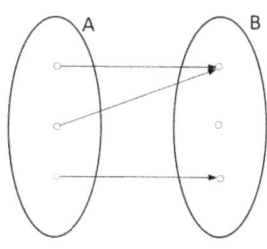

Not onto

Onto functions are also called *surjective*. Note that a function maps its domain *into* its co-domain but its domain *onto* its range.

We will need a more mathematical definition of surjective:

Definition 4.2.2: f is *surjective* iff for every $b \in B$, there is an $x \in A$, such that $f(x)=b$.

4.3 ONE TO ONE FUNCTIONS

Definition 4.3.1: Let f map A into B. f is *one-to-one* (*1-1*) if no two different elements in A have the same image. A *one-to-one* function is also called an *injection*.

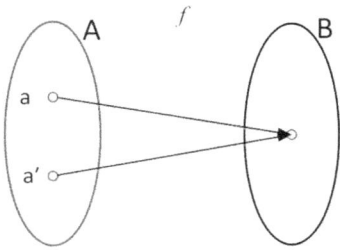

f is not 1-1

Definition 4.3.1 is equivalent to saying that f is one to one, iff for each $y \in Ran\ f$ there is only one $x \in Dom\ f$, such that $f(x)=y$. This leads us to a more mathematical definition of 1-1.

Definition 4.3.1 (a): Let $f : A \to B$. f is *one to one* iff $f(a) = f(a')$ implies that $a = a'$, i.e. if two images are equal then the objects must be equal.

Recall that an implication, $p \to q$, and its contra-positive, $\sim q \to \sim p$, are logically equivalent. Thus we have yet another definition of one to one:

Definition 4.3.1 (b): $f : A \to B$ is *one to one* iff $a \neq a' \to f(a) \neq f(a')$ (i.e. distinct elements have distinct images).

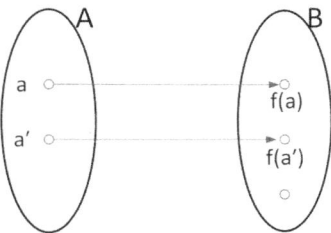

A typical 1-1 function

Example 6:

(i) $f : \Re \to \Re$ defined by $f(x) = x^2$ is not 1-1. We need only find a counter example; for example, $f(2) = 4$ and $f(-2) = 4$; but $2 \neq -2$. Neither is f onto; since the square of a number cannot be negative.

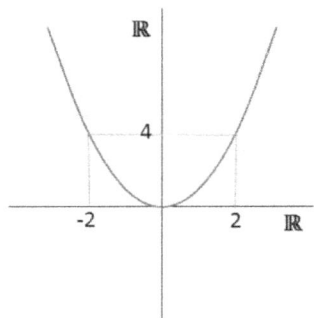

(ii) To obtain a 1-1 function, we must restrict the domain. $f : \Re^+ \to \Re$ such that $x \to x^2$ is 1-1. Every horizontal line cuts the graph *at most* once.

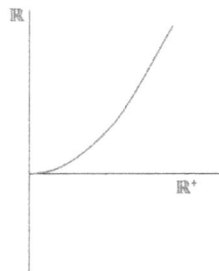

In this case, however, f is not onto. Every horizontal line *must* cut the graph if the function is onto.

(iii) If we take $f : \Re \to \Re^+$, then f is onto but not 1-1.

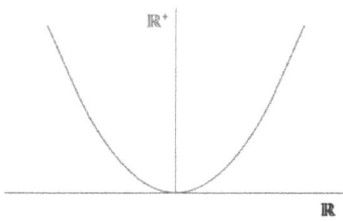

FUNCTIONS

(iv) $f: \Re^+ \to \Re^+$ such that $x \to x^2$ is both 1-1 and onto.

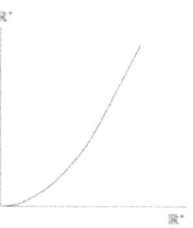

Every horizontal line cuts the graph *once and only once*.

We have been using graphs to illustrate injectivity and surjectivity. The graph however is not a proof. You will learn how to write the proof in the next section. We need just one more definition.

4.3 ONE TO ONE CORRESPONDENCE

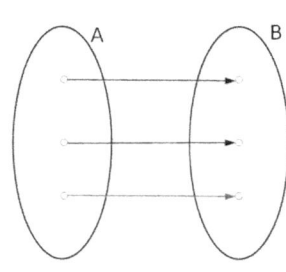

Definition 4.3.1: A function which is both injective and surjective is said to be *bijective*, and is called a *bijection*.

E.g. $f: \Re^+ \to \Re^+$ defined by $f(x) = x^2$ is bijective.

A bijection is also called a *one to one correspondence*. Every element in B has one and only one element in A to which it is assigned.

Example 7: $f: \Re \to \Re$ defined by $f(x) = x^3$ is bijective. For each x^3 there is only one x.

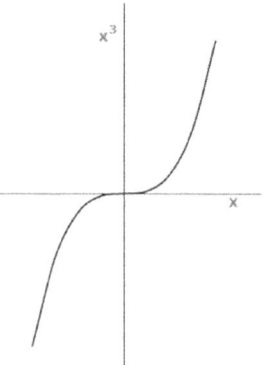

We like our functions to be bijective. They are easier to work with. We go on now to the mathematical proof of bijectivity.

Typical Exam Questions

Example 8:

$f: \Re^+ \to \Re^+$ is such that $f(x) = x^2$. Prove that f is bijective.

Proof:

(i) *One to one*

To show that f is 1-1 we must show that $f(x_1) = f(x_2)$ implies that $x_1 = x_2$.
So, assume that $f(x_1) = f(x_2)$, where $x_1, x_2 \in \Re^+$. Then
$f(x_1) = f(x_2) \Rightarrow x_1^2 = x_2^2$

$\Rightarrow x_1^2 - x_2^2 = 0$

$\Rightarrow (x_1 + x_2)(x_1 - x_2) = 0$

$\Rightarrow x_1 - x_2 = 0$ (since for $x_1, x_2 \in \Re^+$, $x_1 + x_2$ is strictly greater than zero)

$\Rightarrow x_1 = x_2$

$\therefore f$ is one to one.

(ii) *Onto*

To show that f is onto, we must show that for any $b \in \Re^+$ (codomain) there is an $x \in \Re^+$ (domain) such that $f(x) = b$. Let $b \in \Re^+$. We need to produce such an x.

Well, $f(x) = b \Leftrightarrow x^2 = b \Leftrightarrow x = \pm\sqrt{b}$. There are two possibilities: $+\sqrt{b}$ and $-\sqrt{b}$.

Now $+\sqrt{b} \in \Re^+$ for $b \in \Re^+$. We have thus found an x, and need not be concerned with the other possibility.

So for every $b \in \Re^+$, $\exists x \in \Re^+$; namely $x = +\sqrt{b}$, such that $f(x) = b$ [i.e. $\left(\sqrt[+]{b}\right)^2 = b$].

$\therefore f$ is onto.

Since f is both 1-1 and onto, then f is bijective. □

Alternative proof for Part (i):

To show that f is one to one we can show instead that $x_1 \neq x_2 \rightarrow f(x_1) \neq f(x_2)$ or $f(x_1) - f(x_2) \neq 0$. Let us assume that $x_1 \neq x_2 \in \Re^+$. Now, $f(x_1) - f(x_2) = x_1^2 - x_2^2 = (x_1 - x_2)(x_1 + x_2)$. Since $x_1 \neq x_2$; $x_1 - x_2 \neq 0$. Also $x_1 + x_2 > 0$, since $x_1, x_2 \in \Re^+$.

$\therefore f(x_1) - f(x_2) \neq 0$.

$\rightarrow f(x_1) \neq f(x_2)$. $\therefore f$ is 1-1. □

Example 9:

Let $A = \{x : x \geq 2\}$ and $B = \{x : x \geq -4\}$. Let $f : A \rightarrow B$ be such that $f(x) = x^2 - 4x$. Show that f is bijective.

Proof:

(i) *One to One*

Let $x_1 \neq x_2 \in A$. We must show that $f(x_1) \neq f(x_2)$.

$f(x_1) - f(x_2) = x_1^2 - 4x_1 - x_2^2 + 4x_2$

$= (x_1 - x_2)(x_1 + x_2) - 4(x_1 - x_2) = (x_1 - x_2)(x_1 + x_2 - 4).$

Now, $x_1 - x_2 \neq 0$ and $x_1 + x_2 - 4 > 0$ for $x_1, x_2 \geq 2$ and $x_1 \neq x_2$.

$\therefore (x_1 - x_2)(x_1 + x_2 - 4) \neq 0$. Therefore f is 1-1.

(Alternatively, we can show that $f(x_1) = f(x_2) \rightarrow x_1 = x_2$, as follows:

$f(x_1) - f(x_2) = x_1^2 - 4x - x_2^2 + 4x_2 = (x_1 - x_2)(x_1 + x_2 - 4) = 0 \Leftrightarrow x_1 - x_2 = 0$ or $x_1 + x_2 - 4 = 0$, in which case $x_1 = x_2 = 2$, $\therefore x_1 - x_2$ is still zero. $\therefore f$ is 1-1.)

(ii) *Onto*

Let $b \in B$. We must show that $\exists x \in A$, such that $f(x) = b$.

$f(x) = b \Leftrightarrow x^2 - 4x = b \Leftrightarrow x^2 - 4x - b = 0 \Leftrightarrow x = \dfrac{4 \pm \sqrt{16 + 4b}}{2} = 2 \pm \sqrt{4 + b}$.

Now, $2 + \sqrt{4 + b} \geq 2$ for $b \geq -4$. $\therefore 2 + \sqrt{4 + b} \in A$.

So for every $b \in B$, there is an $x \in A$, namely $x = 2 + \sqrt{4 + b}$, such that $f(x) = b$, [i.e. $f(2 + \sqrt{4 + b}) = b$]. Therefore f is onto.

Since f is 1-1 and onto, it is bijective. □

Two Special Functions

Definition 4.3.2: $f : A \rightarrow A$ defined by $f(x) = x$, is called the *identity* function.

The identity function is usually denoted by the letter i. Each element is mapped to itself.

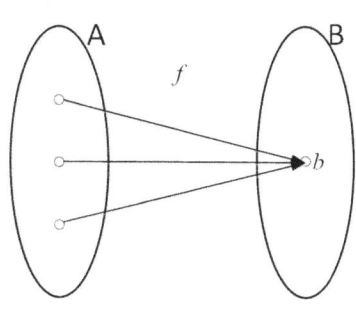

Definition 4.3.3: $f : A \rightarrow B$ defined by $f(x) = b \in B$, is called a *constant* function.

The range of f contains only one element.

Example 10: Let $f : \mathfrak{R} \rightarrow \mathfrak{R}$ be defined by $f(x) = 0$. Then f is a constant function.

In $\mathfrak{R}^2$, the graph of a constant function is a horizontal line.

4.4 COMPOSITION OF FUNCTIONS

Definition 4.4.1: Let $f : A \to B$ and $g : B \to C$. Let $a \in A$ and $g(f(a))$ be the image of $f(a)$ under g. The function $h : A \to C$ defined by $h(a) = g(f(a))$ is called the *composition* of f and g. We write $h = g \circ f$.

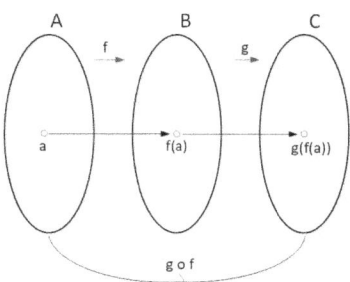

Composition of Functions

Example 11:

Let $A, B, C = \Re$, $f(x) = x^2$ and $g(x) = 2x$, then $(g \circ f)(3) = g(f(3)) = g(9) = 18$. Note, however, that $(f \circ g)(3) = f(g(3)) = f(6) = 36$. Clearly $g \circ f \neq f \circ g$.

Composition of functions is *not* commutative. However, it is associative.

Theorem 4.4.1: Let $f : A \to B$, $g : B \to C$, $h : C \to D$. Then $h \circ (g \circ f) = (h \circ g) \circ f$.

Proof:

$((h \circ g) \circ f)(x) = (h \circ g)(f(x)) = h(g(f(x)))$.

$(h \circ (g \circ f))x = h((g \circ f)(x)) = h(g(f(x)))$.

$\therefore (h \circ (g \circ f))x = ((h \circ g) \circ f)(x)$, for every $x \in A$. □

Exercise: $h : \Re \to \Re$, $h(x) = x^2 + 2$; $g : \Re \to \Re$, $g(x) = x^2$; $f : \Re \to \Re$, $f(x) = x + 2$. Show that $(f \circ g) \circ h = f \circ (g \circ h)$.

Theorem 4.4.2: $\text{Dom}(g \circ f) = \text{Dom } f$, $\text{Ran}(g \circ f) \subseteq \text{Ran } g$.

The example below illustrates Theorem 4.4.2. The proof is left as an exercise.

FUNCTIONS

Example 12:

Let $f : \Re \to \Re$ be such that $f(x) = \sin x$ and $g : \Re \to \Re$ be such that $g(x) = 3x + \pi$.

Then $(g \circ f)(x) = 3 \sin x + \pi$. Clearly $Dom\,(g \circ f) = Dom\,f = \Re$.

We find $Ran\,(g \circ f)$:

$-1 \leq \sin x \leq 1$

$\Rightarrow -3 \leq 3 \sin x \leq 3$

$\Rightarrow \pi - 3 \leq 3 \sin x + \pi \leq \pi + 3$.

$\therefore Ran(g \circ f) = [\pi - 3, \pi + 3]$.

Now $Ran\,g = \Re$, so clearly $Ran(g \circ f) \subseteq Ran\,g$.

Exercise: $f : \Re \to \Re$ is such that $f(x) = x^3 + 7x$, $g : \Re \to \Re$ is such that $g(x) = x + 1$. Find $g \circ f$ and $f \circ g$ and state their range and domain.

Theorem 4.4.3: Let $f : A \to B$ and $g : C \to D$ with $Ran\,f \subseteq C$. If f and g are injective, then so is $g \circ f$.

Proof (by Contradiction)

Suppose $g \circ f$ is *not* injective. Then $\exists\, x_1 \neq x_2 \in A$ such that $g(f(x_1)) = g(f(x_2))$.

Since g is injective, then $g(a) = g(b) \Rightarrow a = b$. $\therefore f(x_1) = f(x_2)$. $\therefore x_1 = x_2$, since f is injective. This is a contradiction. Therefore our assumption is false. $\therefore g \circ f$ is injective. □

4.5 INVERSE FUNCTIONS

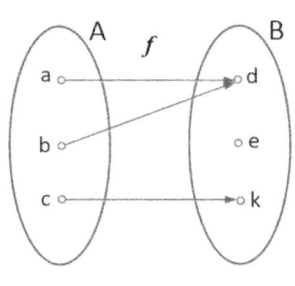

Let $f : A \to B$ and $b \in B$. The inverse of the set element $\{b\}$, denoted $f^{-1}(\{b\})$, is given by $f^{-1}(\{b\}) = \{x : x \in A, f(x) = b\}$.

$f^{-1}(\{b\})$ may or may not be empty.

In the diagram, $f^{-1}(\{d\}) = \{a,b\}$, $f^{-1}(\{k\}) = \{c\}$ and $f^{-1}(\{e\}) = \emptyset$.

If, however, $f : A \to B$ is a one to one correspondence, then for every $b \in B$, $f^{-1}(\{b\})$ will consist of only one element in A. So we have a relation which assigns to each element in B one and only one element in A. We denote this relation by f^{-1}. Then, $f^{-1} : B \to A$ is called the *inverse function* of f.

Example 13:

Let $A = \{x : x \geq 2\}$ and $B = \{x : x \geq -4\}$ and let $f : A \to B$ be such that $f(x) = x^2 - 4x$. We showed in Example 9 that f is bijective and that for every $b \in B$, $f(2 + \sqrt{4+b}) = b$. Therefore $f^{-1} : B \to A$ is such that $f^{-1}(x) = 2 + \sqrt{4+x}$.

More rigorously we have the following definition:

Definition 4.5.1: Let $f : A \to B$ and $g : B \to A$ then g is said to be an *inverse* of f iff $g \circ f = i$ and $f \circ g = i$, that is $f^{-1}(f(x)) = x$, $\forall x \in A$ and $f(f^{-1}(x)) = x$, $\forall x \in B$.

Before we go further, let's take a look at why anyone would want to find the inverse of a function. There are lots of applications. The world would not be same as it is today without inverse functions.

An Application to Engineering

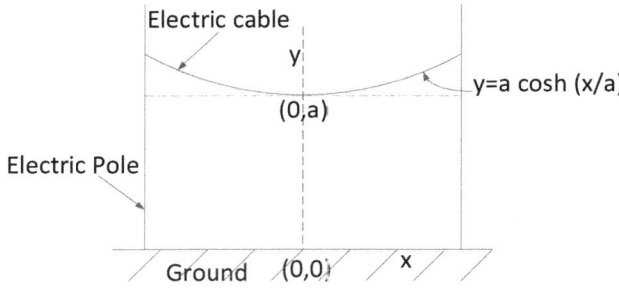

Consider an electrical cable suspended between two poles. If we introduce a coordinate system as shown, the cable forms a curve with equation $y = a\cosh(x/a)$, where a is a constant depending on the tension and physical properties of the wire.

Suppose that for a given value of a and for some height y, the engineer needs to find the distance x from origin. He will need to determine the inverse function.

Theorem 4.5.1: Let $f : A \to B$ then the inverse function, f^{-1}, if it exists, is unique.

Proof:

(To show uniqueness: Suppose that there are two inverses and show that they are equal).

Suppose g and h are inverse functions for $f : A \to B$. Then $f \circ h = i$ and $g \circ f = i$.

$\therefore go(foh) = goi = g$ and $(gof)oh = ioh = h$.

But composition of functions is associative. $\therefore g = h$. The inverse, if it exists, is unique. □

Theorem 4.5.2: Let $f : A \to B$. If f is bijective then f^{-1} exists and is bijective.

Proof: Assume f is bijective. To show f^{-1} exists, we must show that there is a rule, f^{-1}, which assigns to each element in B a unique element in A, such that $f^{-1}(f(x)) = x$. Suppose $\exists t \in B : f^{-1}(t)$ has no image in A. This contradicts f being onto. Suppose that f^{-1} has two images. This contradicts f being 1-1. So the function f^{-1} exists. (Exercise: Show that f^{-1} is bijective). □

NB: We can still define an inverse if f is 1-1 and not onto. If $f : A \to B$ is 1-1 then $f^{-1} : \text{Ran } f \to A$ exists.

[Once bijectiveness is established, finding f^{-1} is easy. It's just like if you put on your socks then put on your shoes. To undo, you must take off your shoes then take off your socks.]

Example 14: $f : \Re \to \Re$ is such that $f(x) = 2x - 3$. We leave it as an exercise to show that f is bijective. To find f^{-1}, we simply solve for x:

$y = 2x - 3 \Rightarrow x = \dfrac{y+3}{2}$. $\therefore f^{-1} : \Re \to \Re$ is such that $f^{-1}(x) = \dfrac{x+3}{2}$.

Example 15: Let $A = \Re - \{2\}$. $f : A \to \Re$ is defined by $f(x) = \dfrac{x+2}{x-2}$.

Is f bijective? Does f^{-1} exist? If so, find f^{-1}.

Solution:

Let $x_1 \neq x_2 \in A$. $f(x_1) - f(x_2) = \dfrac{x_1 + 2}{x_1 - 2} - \dfrac{x_2 + 2}{x_2 - 2}$

$= \dfrac{(x_1 + 2)(x_2 - 2) - (x_1 - 2)(x_2 + 2)}{(x_1 - 2)(x_2 - 2)}$

$= \dfrac{x_1 x_2 + 2x_2 - 2x_1 - 4 - x_1 x_2 - 2x_1 + 2x_2 + 4}{(x_1 - 2)(x_2 - 2)}$

$$= \frac{4(x_2 - x_1)}{(x_1 - 2)(x_2 - 2)}.$$

Since $x_1 \neq x_2$ the numerator of this fraction cannot be zero.

Also $x_1, x_2 \neq 2$, ∴ denominator $\neq 0$ ∴ $f(x_1) - f(x_2) \neq 0$. ∴ f is 1-1.

Now let $b \in \Re$. $f(x) = b \Leftrightarrow \dfrac{x+2}{x-2} = b$. Solving for x, we have:

$$x + 2 = bx - 2b \Leftrightarrow x - bx = -2b - 2 \Leftrightarrow x = \frac{2(b+1)}{b-1}.$$

Is x in A for every $b \in \Re$? Well, if $b = 1$, there is no such x. ∴ f is *not* onto.

$\Rightarrow f$ is not bijective. ∴ $f^{-1}(x): \Re \to A$ is undefined.

We can however define $f^{-1}: \text{Ran } f \to A$. We must exclude $\{1\}$ from the co-domain of f:

$f^{-1}: \Re - \{1\} \to \Re - \{2\}$ is such that $x \to \dfrac{2(x+1)}{x-1}$.

❖❖❖

4.6 FEATURE MATHEMATICIAN

LEONHARD EULER

INVENTOR OF FUNCTIONAL NOTATION

Leonhard Euler (1707-1783) was born in Basel, Switzerland. His genius developed early. A student of Jacob Bernoulli, he began mathematical research at age 18, and is considered probably the most prolific mathematician of all time. Euler's energy and capacity for work were virtually boundless. He made contributions to almost every branch of Mathematics [Anton, 58].

Euler's work on functions demonstrates the importance of language in the development of ideas. Historically, the term "function" was first used by Leibniz in 1673 to denote the dependence of one quantity on another; and following the development of calculus by Leibniz and Newton, results in Mathematics developed rapidly in a disorganized manner. Euler conceived the idea of denoting functions by the letters of the alphabet and his functional notation, *f(x)*, gave coherence to the development of calculus and made way for the more precise definition of a function that we have today.

Euler did much work on Number Theory and Complex Numbers. The most beautiful identity $e^{i\pi} + 1 = 0$ and the number e (= 2.71828...) are named after him. He was the first mathematician to bring the full power of calculus to bear on problems in Physics and he made many contributions in this field as well.

Euler was the son of a Protestant minister and throughout his life was an unaffected and devout Christian. The story is told that when the French philosopher, Denis Diderot, visited Russia, espousing atheism, the empress, Catherine, called upon Euler to confront him with a proof for the existence of God. In a tone of perfect conviction, Euler approached Diderot and announced, "Sir, $\dfrac{a+b^n}{n} = x$, hence God exists—reply!" Diderot stood dumbstruck, and embarrassed by the peals of laughter, took his exit.[1]

What is amazing about Euler is that he was blind for the last 17 years of his life, and this was considered to be his most prolific period. During this time he served as head of the

St. Petersburg Academy in Russia. He had a flawless memory. He once settled an argument between two students by solving a complicated calculation in his head to fifty decimal places! Blindness, or for that matter any physical disability, is no impediment to greatness.

1. "Leonhard Euler," *Wikipedia*, https://en.wikipedia.org/wiki/Leonhard_Euler.

4.7 CHAPTER EXERCISES

1. Let $f : A \to B$ and $g : B \to C$ be two bijective functions. Show that $(gof)^{-1} = f^{-1} o\, g^{-1}$.

2. Let $f : \mathfrak{R}^+ \cup \{0\} \to \mathfrak{R}^+ \cup \{0\}$ be such that $f(x) = x^2$. (i) Show that f is bijective. (ii) Find f^{-1}.

3. Let $f : \mathfrak{R} \to \mathfrak{R}$ be such that $f(x) = x$. (i) Prove that f is bijective. (ii) Find f^{-1}.

4. Let $f : \mathfrak{R}^+ \to \mathfrak{R}$ be defined by $f(x) = 2x - 3$. Determine if f is bijective and find f^{-1}, if possible. Let $g : \mathfrak{R}^+ \to \mathfrak{R}^+$ be such that $g(x) = x^2 + 2$. Find fog and its range.

5. In each case, determine if f is one to one or onto and illustrate f graphically:
(a) $f : \mathfrak{R}^+ \to \mathfrak{R}$ is such that $f(x) = x^2$. (b) $f : \mathfrak{R} \to \mathfrak{R}^+$ is such that $f(x) = x^2$.

6. Let $f : \mathfrak{R} \to \mathfrak{R}$ be such that $f(x) = x^2 + 3$. (i) Show that f is neither one to one nor onto. (ii) Find a restriction g of f which is bijective and has the same range as f. (iii) Find g^{-1}.

7. $f : \mathfrak{R} \to \mathfrak{R}$ is such that $f(x) = x^3$. (i) Prove that f is bijective. (ii) Find f^{-1}.

❖❖❖

CHAPTER REFERENCES

Howard Anton, *Calculus with Analytic Geometry*, 3rd ed. (New York: Wiley, 1988).

J Hunter and D Monk, *Algebra and Number Systems*, (Glasgow: Blackie, 1971).

Seymour Lipschutz, *Schaum's Outline of Theory and Problems of Set Theory and Related Topics, Schaum's Outline Series* (New York : McGraw-Hill, 1977).

CHAPTER FIVE

NATURAL NUMBERS

In this short chapter we give an elementary introduction to the number systems, then focus on the set of Natural Numbers. In particular, we highlight the Principle of Mathematical Induction— a key tool for proving results on natural numbers.

5.1 THE NUMBER SYSTEMS

For the moment, let us just say that real numbers can be represented by points on a straight line. The numbers to the right of 0 are positive numbers and the numbers to the left of 0 are negative. The number 0 is neither positive nor negative.

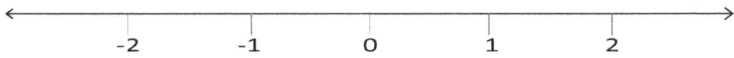

The Real Number Line

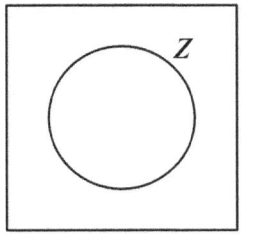

Let us take our universal set to be the set of real numbers $\Re$. We can define the following subsets of $\Re$:

INTEGERS

$Z = \{...,-3,-2,-1, 0, 1, 2,...\}$ is the set of integers or whole numbers. If we let Z^+ be the set of positive integers and Z^- be the set of negative integers, then $Z = Z^+ \cup \{0\} \cup Z^-$.

The set of integers is closed under addition, multiplication and subtraction; but not under division. For example, $\frac{3}{2} \notin Z$.

NATURAL NUMBERS

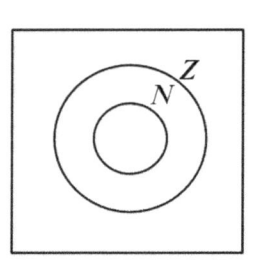

The set of natural numbers, N, consists of zero and the positive integers. Some texts exclude zero from the set of natural numbers; but we will include it in our approach. Thus:

$$N = \{0, 1, 2, 3, \ldots\}.$$

Observe that there is a smallest natural number, namely zero, but there is no smallest integer or real number.

A *prime* number is a natural number, greater than 1, which is only divisible by itself and 1. The natural numbers 0 and 1 are therefore not prime numbers. We will deal with prime numbers in Chapter 7.

RATIONAL NUMBERS

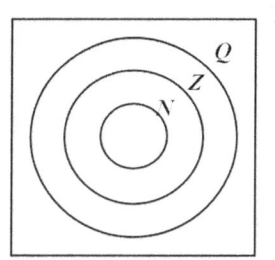

A rational number can be expressed as a ratio of two integers. Let Q denote the set of rational numbers. Then

$$Q = \left\{\frac{m}{n}, \text{where } m \in Z, n \in Z, n \neq 0\right\}.$$

Q is therefore a superset of Z, i.e. $Q \supset Z$. Note that $Q - \{0\}$ is closed under division, but Q is not.

IRRATIONAL NUMBERS

The irrational numbers, Q', is the set of real numbers which are not rational.

We can tell whether a number is rational by looking at its decimal expansion. Every real number can be represented by a non-terminating decimal. For example, $\frac{3}{4} = .75000\ldots$ It is a non-terminating decimal as the zeros repeat. $\frac{22}{7} = 3.\overline{142857142857}$, the last six digits repeat. Irrational numbers have non-repeating decimal expansions. For example,

NATURAL NUMBERS

$\pi = 3.141592...$ The expansion goes on forever and ever without repeating! It is an irrational number. We will learn more about these numbers in Chapter 7.

5.2 INDUCTIVE SETS

Our introduction to number systems merely described the natural numbers as a subset of Z. But what exactly is a natural number? In the nineteenth century Italian mathematician Giuseppe Peano, and others, formulated an axiomatic approach to defining a natural number. It is based on the idea of successor or inductive sets:

Definition 5.2.1: A set S is called an *inductive* set iff (i) $0 \in S$ and (ii) if $x \in S$, then $x+1 \in S$.

Definition 5.2.2: An element n is called a *natural number* iff n is an element of every inductive set.

This may not be the definition that you were expecting. How do we get N from this? Well, by definition of inductive set $0 \in N$, and since $0 \in N$ then $0+1 \in N$. Therefore $1 \in N$. But if n is a natural number then $n+1$ is also a natural number. $\therefore N = \{0, 1, 2, 3, ...\}$. These are the natural numbers that we all know and love.

Now that we have our definition of N, we will use it in later chapters to construct larger sets from N (namely, Z, Q and $\Re$) and to develop our number theory. We can also use the definition to prove results on natural numbers. Because the Principle of Induction is such a vital tool for proof in Mathematics, we devote the rest of the chapter to it.

5.3 PRINCIPLE OF MATHEMATICAL INDUCTION

Theorem 5.3.1: Let $S \subseteq N$ be an inductive set, then $S = N$.

Proof: We already know that $N \subseteq S$ (each natural number is an element of every inductive set). Since $S \subseteq N$ by assumption, then $S = N$. □

Theorem 5.3.1 essentially says that the set of natural numbers is the smallest inductive set. There exists larger inductive sets (also called successor sets). For example, the set Z is inductive since $0 \in Z$ and if $z \in Z$, then $z+1 \in Z$. You will meet these up in Second Year Mathematics. For the moment, we are interested in the smallest inductive set.

Principle of Mathematical Induction:

If $S \subseteq N$ is such that $0 \in S$ and $n+1 \in S$ whenever $n \in S$, then $S = N$.

The Principle of Mathematical Induction is clearly the same as Theorem 5.3.1. We have already proven it! Here it is again in a more practical form.

Principle of Mathematical Induction:

Let $T(n)$ be a result corresponding to the natural numbers and suppose that (i) $T(0)$ is true and (ii) for every natural number k, the truth of $T(k)$ implies the truth of $T(k+1)$. Then $T(n)$ is true for every natural number n.

Proof:

Define the set S by $S = \{n : n \in N \text{ and } T(n) \text{ is true}\}$. Then $S \subseteq N$. But S is an inductive set since $0 \in S$ and $k \in S$ implies $k+1 \in S$. $\therefore S = N$. Therefore $T(n)$) is true for all natural numbers. □

Grand conclusion: To show that a result is true for all natural numbers—i.e. all the elements of $N = \{0, 1, 2, ..., k, k+1, ...\}$, first show that it is true for 0; then show that if it is true for any natural number k, then it is true for $k+1$. Then, by the Principle of Mathematical Induction, it is true for all $n \in N$.

We can also prove a result holds for certain infinite subsets of N, by beginning with the truth of the smallest element of the subset.

Example 1(a): Prove that $0 + 1 + 2 + 3 + 4 + ... + n = \dfrac{n(n+1)}{2}$, for any $n \in N$.

Proof (induction on n):

(i) Let $n = 0$, L.H.S. $= 0$, R.H.S $= 0$. Therefore the result is true for $n = 0$.

(ii) Suppose that the result is true for $n = k$, where k is some natural number.

i.e. $0 + 1 + ... + k = \dfrac{k(k+1)}{2}$.

Then $0 + 1 + ... + k + k + 1 = \dfrac{k(k+1)}{2} + k + 1$

$= \dfrac{k(k+1) + 2(k+1)}{2} = \dfrac{(k+1)(k+2)}{2}$

The result is therefore true for $n = k+1$, if true for $n = k$. Since it is also true for $n = 0$, then by the Principle of Mathematical Induction it is true for all natural numbers n. □

This method is better than trying to prove the theorem by substituting each natural number in turn into both sides of the equation. Your proof would take forever and ever this way!

Employing summation notation can help us to write a more elegant proof:

SUMMATION NOTATION

(i) $\sum_{i=1}^{k} a_i = a_1 + a_2 + ... + a_k$. i is called the summation index, k is the upper limit of the summation.

E.g. $\sum_{i=1}^{10} i^2 = 1^2 + 2^2 + ... + 10^2 = \sum_{r=1}^{10} r^2$. (N.B. i and r are 'dummy' variables. It does not matter which letter we use.)

(ii) $\sum_{i=1}^{k+1} a_i = a_1 + a_2 + ... + a_k + a_{k+1} = \sum_{i=1}^{k} a_i + a_{k+1}$

(iii) $\sum (a_i + b_i) = \sum a_i + \sum b_i$

(iv) $\sum c a_i = c \sum a_i$, where c is a constant.

In Example 1(b) below, we use summation notation to redo Example 1(a). For illustrative purposes we prove the result for positive natural numbers instead.

Example 1(b): Prove $\sum_{i=1}^{n} i = \dfrac{n(n+1)}{2}$, for all non-zero natural numbers n.

Proof (induction on n):

Let $T(n) : \sum_{i=1}^{n} i = \dfrac{n(n+1)}{2}$

First we establish truth of $T(1)$: $\sum_{i=1}^{1} i = 1 = \dfrac{1(1+1)}{2}$. ∴ $T(1)$ is true.

Next we assume the truth of $T(k)$, i.e. we assume that $\sum_{i=1}^{k} i = \dfrac{k(k+1)}{2}$ for some non-zero natural number k.

Then we show that $T(k+1)$ is true:

$$\sum_{i=1}^{k+1} i = \sum_{i=1}^{k} i + (k+1) = \frac{k(k+1)}{2} + k + 1 \text{ (by assumption)}$$

$$= \frac{k(k+1) + 2(k+1)}{2} = \frac{(k+1)(k+2)}{2}.$$

$\therefore T(k) \to T(k+1)$.

$\therefore$ By the Principle of Mathematical Induction, $T(n)$ is true for all non-zero natural numbers n. □

Example 2: Prove that for all $n \in Z^+$

$$\sum_{r=1}^{n} r^2 = \frac{n}{6}(n+1)(2n+1)$$

Proof (induction on n):

The theorem is true for $n=1$, since $1^2 = \frac{1}{6}(2)(3)$.

Now assume true it is true for $n=k$, i.e. $\sum_{r=1}^{k} r^2 = \frac{k}{6}(k+1)(2k+1)$.

We must show that $\sum_{r=1}^{k+1} r^2 = \frac{(k+1)}{6}(k+2)(2(k+1)+1)$:

Well, $\sum_{r=1}^{k+1} r^2 = \frac{(k)}{6}(k+1)(2k+1) + (k+1)^2$

$$= \frac{(k+1)[k(2k+1) + 6(k+1)]}{6} = \frac{(k+1)}{6}[2k^2 + 7k + 6]$$

$$= \frac{(k+1)}{6}(k+2)(2k+3) = \frac{(k+1)}{6}(k+2)[2(k+1)+1]$$

$\therefore T(k) \to T(k+1)$.

By the Principle of Mathematical Induction the result is true for all positive integers. □

NATURAL NUMBERS

We can use induction to prove just about any result on natural numbers. Here is another type of theorem.

Example 3: Prove that for any positive integer n,
$$3 \cdot 5^{2n+1} + 2^{3n+1} \text{ is exactly divisible by } 17.$$

Proof (induction on n):

The statement is true for $n = 1$ since $3 \cdot 5^3 + 2^4 = 391$ — this is divisible by 17.

Assume that $3 \cdot 5^{2k+1} + 2^{3k+1}$ is divisible by 17, for some positive integer k.

Then $3 \cdot 5^{2(k+1)+1} + 2^{3(k+1)+1} = 3 \cdot 5^{2k+1} \cdot 5^2 + 2^{3k+1} \cdot 2^3$

$$= 17 \cdot 3 \cdot 5^{2k+1} + 8 \cdot 3 \cdot 5^{2k+1} + 8 \cdot 2^{3k+1}$$

$$= 17 \cdot 3 \cdot 5^{2k+1} + 8 \left(3 \cdot 5^{2k+1} + 2^{3k+1} \right)$$

The sum of two numbers which are both divisible by 17 is clearly divisible by 17. Therefore the result is true for $n = k+1$, if true for $n = k$.

$\therefore$ By the Principle of Mathematical Induction the result is true for all $n \in Z^+$. $\square$

Example 4: Prove for any $n \in N$
$$3^{(4n+2)} + 5^{(2n+1)} \text{ is exactly divisible by } 14.$$

Proof (induction on n):

The result is true for $n = 0$ since $3^2 + 5 = 14$, which is divisible by 14.

Now assume that $3^{4k+2} + 5^{2k+1}$ is divisible by 14, for some natural number k.

Then $3^{4(k+1)+2} + 5^{2(k+1)+1} = 3^{4k+2} \cdot 3^4 + 5^{2k+1} \cdot 5^2$

$$= 3^{4k+2}(25 + 56) + 5^{2k+1} \cdot 25 \text{, since } 3^4 = 81 = (5^2 + 56).$$

$$= 25 \left(3^{4k+2} + 5^{2k+1} \right) + 14 \cdot 4 \cdot 3^{4k+2}.$$

The first term of the above quantity is divisible by 14 by assumption and the second term is clearly divisible by 14. Therefore the sum of the two terms is divisible by 14. Therefore the theorem is true for $n = k+1$ if true for $n = k$. Since it is also true for $n = 0$, by the Principle of Mathematical Induction it is true for all $n \in N$. $\square$

After you pick up the form of the proof you will be able to handle more challenging examples.

Example 5: Prove that for all $n \in Z^+$

$$\sum_{r=1}^{n} \frac{1}{r(r+1)(r+2)} = \frac{n(n+3)}{4(n+1)(n+2)}$$

Proof:

When $n = 1$, L.H.S. $= \dfrac{1}{1 \cdot 2 \cdot 3} = \dfrac{1 \cdot 4}{4 \cdot 2 \cdot 3} = \dfrac{1}{6} =$ R.H.S. Hence the result holds for $n = 1$.

Now assume that it is true for $n = k$, i.e.

$$\sum_{r=1}^{k} \frac{1}{k(k+1)(k+2)} = \frac{k(k+3)}{4(k+1)(k+2)}$$

Then $\displaystyle\sum_{r=1}^{k+1} \frac{1}{r(r+1)(r+2)} = \sum_{r=1}^{k} \frac{1}{r(r+1)(r+2)} + \frac{1}{(k+1)(k+2)(k+3)}$

$= \dfrac{k(k+3)}{4(k+1)(k+2)} + \dfrac{1}{(k+1)(k+2)(k+3)} = \dfrac{1[k(k+3)^2 + 4]}{4(k+1)(k+2)(k+3)}$

$= \dfrac{k(k^2 + 6k + 9) + 4}{4(k+1)(k+2)(k+3)} = \dfrac{k^3 + 6k^2 + 9k + 4}{4(k+1)(k+2)(k+3)} = \dfrac{(k+1)(k+4)(k+1)}{4(k+1)(k+2)(k+3)}$

$= \dfrac{(k+1)(k+4)}{4(k+2)(k+3)}$

$\therefore$ The result is true for $n = k+1$.

By the Principle of Mathematical Induction the result is true for all $n \in Z^+$. $\square$

5.4 FEATURE MATHEMATICIAN

GIUSEPPE PEANO

"WHAT ARE NUMBERS?"

Giuseppe Peano
(1858-1932)

We have been using natural numbers all our lives. Even from babies, we learn to count and put things in order. But what exactly is a natural number? Italian mathematician Giuseppe Peano was among the first to define the set of Natural Numbers.

Peano was born in Italy in 1858. The son of a poor farmer, he obtained his early education from the village school. Recognising his talent, his uncle took him to Turin, when he was 12 years old. He later entered Turin University intending to pursue Engineering; but switched to Mathematics. He obtained his Ph.D. in 1880. Peano stayed on as staff at Turin.

In 1889, he postulated an axiomatic foundation for the set of natural numbers based on successor (inductive) sets. The following are the Peano Axioms:

1. Zero is a number.
2. The successor of any number is another number.
3. There are no two numbers with the same successor.
4. Zero is not the successor of any number.
5. If a set S of numbers contains zero and also the successor of every number in S, then every number is in S.

You will recognize Axiom 5 as the Principle of Mathematical Induction, as stated in this chapter. Peano was not the first to use induction. Implicit forms of induction are found in the writings of Euclid (300 BC). French mathematician, Blaise Pascal (1665), was first to give a clear statement of Induction; but German mathematician, Richard Dedekind (1887), is usually credited as its inventor. Peano himself attributed the formulation of his axioms to Dedekind, though he developed and popularized them. The whole number system that you are so familiar with, and everything that you learnt about the arithmetic of numbers, can be derived from the Peano Axioms.

Peano made many other contributions to Mathematics. He died in 1932 while still a teacher at Turin. You will meet his axioms again in Year 2 Mathematics.

5.5 CHAPTER EXERCISES

1. Prove by induction the following (n is a natural number)

(i) $\sum_{r=1}^{n} r(r+1)^2 = n(n+1)(n+2)(3n+5)/12$

(ii) $7^n + 2$ is divisible by 3   (iii) $11^n - 6$ is divisible by 5.

2. Prove that if $x \geq 0$ and n is a positive integer, then $(1+x)^n \geq 1+nx$.

3. Prove that $n! \leq n^n$ for all positive integers n.

4. Prove that for all positive integers n

$$\sum_{r=1}^{n} r^3 = \frac{n^2(n+1)^2}{4}.$$

5. Prove that for all $n \in N$,

$$\sum_{r=1}^{n} \frac{r}{r^4 + r^2 + 1} = \frac{n(n+1)}{2(n^2+n+1)}.$$

◇◇◇

CHAPTER REFERENCES

Herbert B. Enderton, *A Mathematical Introduction to Logic*, 2 ed. (New York: Academic/Elsevier, 2006).

P.A. Morris, *Introduction to Algebra*, 3rd edition (Department of Mathematics, UWI, St Augustine, 1996).

Seymour Lipschutz, *Schaum's Outline of Theory and Problems of Set Theory and Related Topics, Schaum's Outline Series* (New York : McGraw-Hill, 1977).

CHAPTER SIX

REAL NUMBERS

In this chapter we look at some elementary topics in the study of real numbers. We first introduce the concept of order and give proof of some standard real number inequalities. We then define the absolute value of a real number and prove the basic theorems on absolute value. We also review techniques for solving real number inequalities, including inequalities involving absolute value. Continuing from our study of the natural numbers in Chapter 5, you may prefer to go first to Chapter 7 and do some study of integers and rational numbers; but we present the real number topics here as students are already familiar with them.

6.1 THE 'LESS THAN' RELATION

Definition 6.1.1: The real number a is *less than* real number b (written $a < b$) iff $b - a$ is a positive number.

Geometrically a is to the left of b on the number line.

The less than relation introduces the concept of order in the real numbers. The following properties can be easily derived:

Order Properties of Real Numbers

Let a, b, c, be real numbers

1. Either $a < b$, $a = b$ or $b < a$. (Law of trichotomy- only one of the three must hold).
2. If $a < b$ and $b < c$, then $a < c$ (transitive property).
3. If $a < b$, then $a + c < b + c$.
4. (i) If $a < b$ and c is a positive number, then $ac < bc$.

    (ii) If $a < b$ and c is a negative number, then $bc < ac$.

Properties 1 and 2 above make $\Re$ an ordered set. The set N is also an ordered set and furthermore it is a well ordered set. This means we can write the elements in some order. One obvious ordering is: 0, 1, 2, 3 . . . . We can do this for Z also, for example: 0, -1, 1, -2, 2, -3, 3 . . . . This is not the usual less than ordering but it is a well-ordering. We can even do it for Q. Chapter 7 will give you a hint. But can you think of a well-ordering for $\Re$? You will become rich and famous overnight.

Archimedian Property of Real Numbers

Recall that the set N of natural numbers has a smallest element; namely zero. So zero is a lower bound for N. However, N is not bounded above. This is equivalent to saying that for each positive real number x, there exists a positive integer n such that $x < n$. This brilliant statement may seem obvious to you now; but it was first claimed 2200 years ago by Greek mathematician, Archimedes. We feature him at the end of the chapter as one of the prime founders of Mathematics. The Archimedean property is important. From it, we can show that there is a rational number between any two real numbers and prove the existence of irrational numbers. But this is Year 2 Math—bet you can't wait!

6.2 ABSOLUTE VALUE

Definition 6.2.1: Let $x \in \Re$. The *absolute value* (or *modulus*) of x, denoted $|x|$, is given by

$$|x| = \begin{cases} x & \text{if } x \geq 0 \\ -x & \text{if } x < 0. \end{cases}$$

Example 1: $|3| = 3$ and $|-3| = -(-3) = 3$.

REAL NUMBERS

Geometrically, the absolute value of x is the distance (positive) between x and the origin.

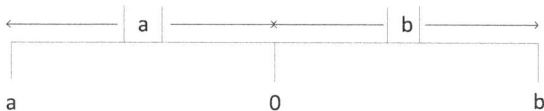

Properties of Absolute Value

1. $|x| \geq 0, \ \forall x \in \Re$ (The absolute value is always non negative).

2. $|x| = 0, \ \text{iff } x = 0$.

Distance between Two Real Numbers

Let a and b be real numbers. If a is less than b, the distance between a and b is $b - a$ = $|b-a|$ = distance between b and a. If b is less than a, the distance between b and a is $a - b = |a - b|$ = distance between a and b. Clearly then $|b - a| = |a - b|$ = distance between a and b, regardless of which is bigger.

Theorems on Absolute Value

We now prove some key theorems on absolute value. Since absolute value is defined piecewise we usually need to consider two or more cases with proofs involving absolute value.

Theorem 6.2.1: Let $y \geq 0$ be a fixed real number, then

$$|x| \leq y, \ \text{iff} - y \leq x \leq y.$$

Proof:
We assume throughout that $y (\geq 0)$ is a fixed real number.

We must prove: (1) $|x| \leq y \rightarrow -y \leq x \leq y$ and (2) $-y \leq x \leq y \rightarrow |x| \leq y$.

Proof of (1). We consider two cases: (i) $x \geq 0$ and (ii) $x < 0$. In both cases, we assume that $|x| \leq y$ and show that $-y \leq x \leq y$.

Case (i) If $x \geq 0$, then $|x| = x$. Hence $|x| \leq y \rightarrow x \leq y$. Clearly $-y \leq x$. $\therefore$ $-y \leq x \leq y$.

Case (ii) If $x < 0$, then $|x| = -x$. Hence $|x| \leq y \rightarrow -x \leq y \rightarrow x \geq -y$ or $-y \leq x$. Clearly also $x \leq y$, since x is negative and $y \geq 0$. $\therefore$ $-y \leq x \leq y$.

In both cases we have $|x| \leq y \rightarrow -y \leq x \leq y$.

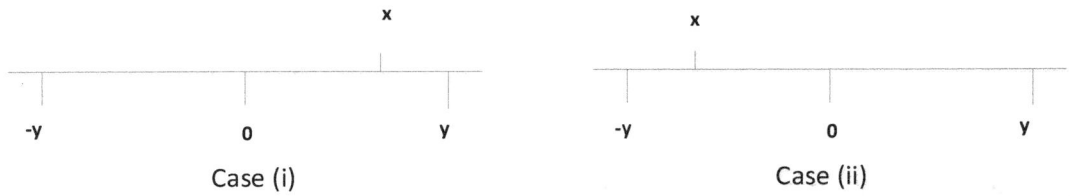

Case (i) Case (ii)

Proof of (2). Conversely, we show that $-y \leq x \leq y \rightarrow |x| \leq y$.

Now assume that $-y \leq x \leq y$, where $y \geq 0$ is a fixed real number.

Case (i) $x \geq 0$. $\therefore$ $|x| = x$

$\therefore -y \leq x \leq y \Rightarrow 0 \leq x \leq y \Rightarrow 0 \leq |x| \leq y$ $\therefore$ $|x| \leq y$.

Case (ii). Now suppose that $x < 0$. Then $|x| = -x$.

$\therefore -y \leq x \leq y \Rightarrow -y \leq x < 0 \Rightarrow y \geq -x \Rightarrow y \geq |x|$. $\therefore$ $|x| \leq y$.

In either case, $-y \leq x \leq y \Rightarrow |x| \leq y$.

From (1) and (2) we have $|x| \leq y$ iff $-y \leq x \leq y$. □

Example 2: $|x| \leq 5 \Leftrightarrow -5 \leq x \leq 5$.

Theorem 6.2.2:

(a). $|-x| = |x|$ (b). $|xy| = |x||y|$ (c). $-|x| \leq x \leq |x|$

Proof of (a)

$$|-x| = \begin{cases} -x & \text{if } -x \geq 0 \\ -(-x) & \text{if } -x < 0 \end{cases} = \begin{cases} -x & \text{if } x \leq 0 \\ x & \text{if } x > 0 \end{cases} = |x|. \square$$

REAL NUMBERS

Proof of (b): We consider four cases:

Case (i). Suppose $x \geq 0$ and $y \geq 0$. Then $xy \geq 0$. $\therefore |xy| = xy = |x||y|$.

Case (ii) If $x \leq 0$ and $y \leq 0$, then $xy \geq 0$. $\therefore |xy| = xy = (-x)(-y) = |x||y|$.

Case (iii) If $x \leq 0$ and $y \geq 0$, then $xy \leq 0$. $\therefore |xy| = -xy = |x||y|$.

Case (iv) is similar to Case (iii). □

Proof of (c) $-|x| \leq x \leq |x|$:

If $x \geq 0$, then $x = |x|$; and clearly $x \geq -|x|$. $\therefore -|x| \leq x = |x|$. $\therefore -|x| \leq x \leq |x|$.

If $x < 0$ then $x = -|x|$; and clearly $x < |x|$. $\therefore -|x| = x \leq |x|$. $\therefore -|x| \leq x \leq |x|$. □

Example 3: $-|3| \leq 3 \leq |3|$ and $-|-3| \leq -3 \leq |-3|$. In each case there is equality on one side.

Triangle Inequality

Theorem 6.2.3: For all $x, y \in \Re$, $|x + y| \leq |x| + |y|$.

(i.e. the absolute value of a sum is always less than or equal to the sum of the absolute values.)

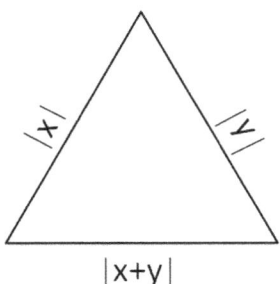

$|x+y|$

The triangle inequality also applies to vectors and tells us that the sum of the lengths of any two sides of a triangle must be greater than or equal to the length of the third side.

Proof: Let $x, y \in \Re$. From Theorem 6.2.2(c) we have $-|x| \leq x \leq |x|$ and $-|y| \leq y \leq |y|$.

Adding these inequalities, we get $-|x| - |y| \leq x + y \leq |x| + |y|$

$\Rightarrow -(|x| + |y|) \leq x + y \leq |x| + |y|$.

$\Rightarrow |x + y| \leq |x| + |y|$ since $-b \leq a \leq b \Rightarrow |a| \leq b$. □

Corollary: $|x - y| \leq |x| + |y|$.

Proof: The result follows immediately from the triangle inequality. If we let $y = -y$ then $|x + -y| \leq |x| + |-y| = |x| + |y|$. □

Theorem 6.2.4: (i) $||x| - |y|| \leq |x + y|$ (ii) $||x| - |y|| \leq |x - y|$.

Proof:

(i) $|x| = |x + y - y| \leq |x + y| + |y|$.

$\therefore |x| - |y| \leq |x + y|$. (1)

Also $|y| = |x + y - x| \leq |x + y| + |x|$. $\therefore |y| - |x| \leq |x + y|$.

$\therefore -(|x| - |y|) \leq |x + y|$ (2)

From (1) and (2), we have $||x| - |y|| \leq |x + y|$.

(ii) Substitute $y = -y$ in (i) to get $||x| - |y|| \leq |x - y|$. □

Example 4: Prove $|y_1| - |y_2| \leq |y_1 - y_2|$, $\forall\ y_1, y_2 \in \Re$.

Proof: Let $x_1 = y_1 - y_2$ and $x_2 = y_2$, where $x_1, x_2 \in \Re$.

From Δ inequality, $|x_1 + x_2| \leq |x_1| + |x_2|$. $\therefore |y_1| \leq |y_1 - y_2| + |y_2|$.

$\therefore |y_1| - |y_2| \leq |y_1 - y_2|$. □

6.3 PROVING REAL NUMBER INEQUALITIES

Some Standard Inequalities

The most fundamental inequality of real numbers is:

$$x^2 \geq 0, \text{ for all real } x.$$

From this basic inequality we can prove others.

Theorem 6.3.1: If a and b are real numbers, then, $a^2 + b^2 \geq 2ab$; with equality iff $a = b$.

Proof:

Begin with $(a-b)^2 \geq 0$. $\therefore$ $a^2 + b^2 - 2ab \geq 0$. $\therefore$ $a^2 + b^2 \geq 2ab$.

Also $(a-b)^2 = 0$ iff $a - b = 0$. Therefore equality holds iff $a = b$. $\square$

Note that we cannot begin with what we want to prove. To prove that $a^2 + b^2 \geq 2ab$, if we say: $a^2 + b^2 \geq 2ab \rightarrow a^2 + b^2 - 2ab \geq 0 \rightarrow (a-b)^2 \geq 0$, we have not proven what we were asked to prove. (In proving *any* result, never assume what you are asked to prove.)

Definition 6.3.1: The *arithmetic mean* of two positive real numbers x and y is $\dfrac{x+y}{2}$.

The *geometric mean* is $\sqrt{xy}$.

Note that $\sqrt{x}$ means positive square root, unless otherwise stated.

Theorem 6.3.2: $\dfrac{x+y}{2} \geq \sqrt{xy}$, for all $x, y \geq 0$.

Proof: Let $a, b \in \Re$. Then $(a-b)^2 \geq 0 \Rightarrow a^2 + b^2 \geq 2ab$. Let $a^2 = x$ and $b^2 = y$, then $x + y \geq 2\sqrt{xy} \Rightarrow \dfrac{x+y}{2} \geq \sqrt{xy}$. $\square$

Thus for any pair of positive real numbers, the arithmetic mean is always greater than or equal to geometric mean.

Example 5:

Prove that if a, b, c are unequal and positive then $(a+b)(b+c)(c+a) \geq 8abc$. When does equality hold?

Proof:

We know $(a+b) \geq 2\sqrt{ab}$, $(b+c) \geq 2\sqrt{bc}$ and $(c+a) \geq 2\sqrt{ca}$.

$\therefore (a+b)(b+c)(c+a) \geq 2\sqrt{ab} \, 2\sqrt{bc} \, 2\sqrt{ac} = 8\sqrt{a^2 b^2 c^2}$.

$\therefore (a+b)(b+c)(c+a) \geq 8abc$. $\square$

Equality holds only if $a = b = c$.

Example 6: Prove that $\forall a,b,c \in \Re$, $a^2 + 2b^2 + c^2 \geq 2b(a+c)$.

Proof:

From Theorem 6.3.1, $a^2 + b^2 \geq 2ab$, $\forall a,b \in \Re$. Similarly, $c^2 + b^2 \geq 2cb$, $\forall b,c \in \Re$.

Adding, we get $a^2 + b^2 + b^2 + c^2 \geq 2ab + 2bc$, $\forall a,b,c \in \Re$.

$\therefore$ $a^2 + 2b^2 + c^2 \geq 2b(a+c)$, $\forall a,b,c \in \Re$. $\square$

Example 7: Prove that for all $a,b,c \in \Re^+$, where $a > b$,

$$1 < \frac{a+c}{b+c} < \frac{a}{b}.$$

Proof:

We must show that $1 < \frac{a+c}{b+c}$ and $\frac{a+c}{b+c} < \frac{a}{b}$.

Clearly $a+c > b+c$ since $a > b$ and $c > 0$. Therefore $1 < \frac{a+c}{b+c}$.

Now $\frac{a+c}{b+c} < \frac{a}{b} \Leftrightarrow (a+c)b < (b+c)a \Leftrightarrow cb < ca \Leftrightarrow b < a$. This is given.

$\therefore$ $\frac{a+c}{b+c} < \frac{a}{b}$. $\square$

Alternatively, suppose that $\frac{a+c}{b+c} \geq \frac{a}{b}$, then $\frac{a+c}{b+c} - \frac{a}{b} \geq 0$.

$\Rightarrow \frac{(a+c)b - a(b+c)}{b(b+c)} \geq 0 \Rightarrow \frac{bc - ac}{b(b+c)} \geq 0 \Rightarrow \frac{c(b-a)}{b(b+c)} \geq 0$.

Since $a,b,c > 0$ then $c(b-a) \geq 0$. $\therefore$ $b \geq a$.

This is a contradiction, since it is known that $b < a$. $\therefore$ $\frac{a+c}{b+c} < \frac{a}{b}$. $\square$

REAL NUMBERS

6.4 SOLVING REAL NUMBER INEQUALITIES

The reader should already be familiar with techniques for solving simple inequalities and expressing the solution in interval notation. Recall that for open and closed sets a parenthesis indicates that the end point is not included in the interval, while a square bracket indicates that the number is included.

Example 8:

(i) $\{x : 2 < x \leq 5\}$ can be written as the half open (or half closed) interval $(2,5]$. The diagram below shows this interval on the number line.

(ii) $\{x : x > 1\}$ can be written as $(1, \infty)$. It is an infinite interval.

We will make use of the following theorem, when solving inequalities:

Theorem 6.4.1:

(i) $ab > 0$ iff $a > 0$ and $b > 0$ or if $a < 0$ and $b < 0$.

(ii) $ab < 0$ iff $a > 0$ and $b < 0$ or if $a < 0$ and $b > 0$.

Similarly for a/b.

Example 9:. Solve $(x + 2)(x - 5) > 0$.

Solution:

There are two cases: either (i) $x + 2 > 0$ and $x - 5 > 0$ or (ii) $x + 2 < 0$ and $x - 5 < 0$.

(i) In this case, $x > -2$ and $x > 5 \Rightarrow x > 5$.

(ii) In this case, $x < -2$ and $x < 5 \Rightarrow x < -2$.

The solution set is $S = \{x : x > 5 \text{ or } x < -2\}$. We can also write $S = (-\infty, -2) \cup (5, \infty)$.—Answer.

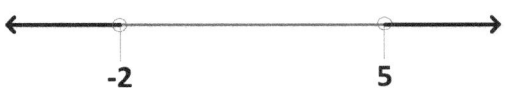

One can also solve the inequality by solving the corresponding equation and testing points:

Example 9 (alternative solution):

First solve $(x+2)(x-5) = 0$. The roots are $x = -2$ and $x = 5$. Test points to see that $(x+2)(x-5)$ is positive for $x \in (-\infty, -2)$, negative for $x \in (-2, 5)$ and positive for $x \in (5, \infty)$. Thus $(x+2)(x-5) > 0$, for $x < -2$ or $x > 5$.

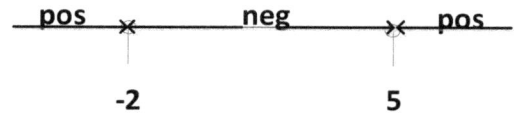

In general, $(x-a)(x-b) \geq 0$; unless $a < x < b$.

Example 10: Solve $\dfrac{2x-5}{x-2} < 1$, $x \neq 2$.

Solution: First put the inequality in the form $\dfrac{a}{b} < 0$.

$$\frac{2x-5}{x-2} - 1 < 0 \Rightarrow \frac{2x-5-(x-2)}{x-2} < 0 \Rightarrow \frac{x-3}{x-2} < 0.$$

Either (i) $x - 3 > 0$ and $x - 2 < 0$ or (ii) $x - 3 < 0$ and $x - 2 > 0$.

(i) $\Rightarrow x > 3$ and $x < 2$, in which case, there is no solution.

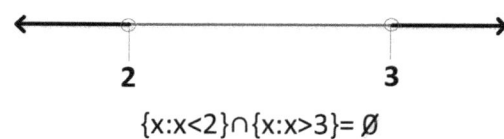

(ii) $\Rightarrow x < 3$ and $x > 2$, in which case, the solution is $2 < x < 3$.

The solution set to the given problem is therefore $(2, 3)$. —Answer.

Example 11: Find the values of x which satisfy $\left\{x \in \Re : \dfrac{x-2}{x} > \dfrac{x}{x+2}, x \neq 0, -2\right\}$.

Solution: $\dfrac{x-2}{x} > \dfrac{x}{x+2} \quad x \neq 0, -2$

$\Rightarrow \dfrac{x-2}{x} - \dfrac{x}{x+2} > 0$

$\Rightarrow \dfrac{(x-2)(x+2) - x^2}{x(x+2)} > 0$

$\Rightarrow \dfrac{-4}{x(x+2)} > 0.$

Since the numerator is negative the denominator must also be negative. There are two cases:

(i) $x > 0$ and $x + 2 < 0$. This is impossible.

(ii) $x < 0$ and $x + 2 > 0 \Rightarrow -2 < x < 0$.

The solution set is therefore $\{x : -2 < x < 0\}$.—Answer.

6.5 INEQUALITIES INVOLVING ABSOLUTE VALUE

Theorem 6.5.1: $|x|^2 = x^2$.

Proof: Recall that $|x| = \begin{cases} x & \text{if } x \geq 0 \\ -x & \text{if } x < 0 \end{cases}$. Since $(-x)^2 = x^2$, we have that $|x|^2 = x^2$, for all x. □

Taking the square root on both sides gives us an alternative definition for absolute value.

Definition 6.5.1: $\sqrt{x^2} = |x|$.

It is a mistake to write $\sqrt{x^2} = x$. Rather, we have $\sqrt{x^2} = \begin{cases} x & \text{if } x > 0 \\ -x & \text{if } x \leq 0. \end{cases}$

For example, suppose that $x = -3$. Then $\sqrt{x^2} = \sqrt{(-3)^2} = \sqrt{9} = 3 = -(x) \neq x$.

These facts will be useful in solving inequalities involving absolute value.

Example 12: Solve $|x-1| > 2|x-2|$.

Solution: Squaring both sides we have $|x-1|^2 > (2|x-2|)^2$

$\therefore (x-1)^2 > 4(x-2)^2$ (since $|x|^2 = x^2$)

$\therefore x^2 + 1 - 2x > 4(x^2 + 4 - 4x)$

$\therefore 3x^2 + 15 - 14x < 0$

$\therefore (3x-5)(x-3) < 0$

Solution $\left\{x : \dfrac{5}{3} < x < 3\right\}$. —Answer.

Example 13: Find the values of x which satisfy $\{x \in \Re : |x+3| < 5\} \cap \{x \in \Re : x^3 - 3x^2 \leq 10x\}$.

Solution: $|x+3| < 5 \to -5 < x+3 < 5 \to -8 < x < 2$.

Also $x^3 - 3x^2 - 10x \leq 0 \to x(x^2 - 3x - 10) \leq 0 \to x(x-5)(x+2) \leq 0$.

The critical points are $x = -2, 0$ and 5. We test points in the four intervals:

	x	x-5	x+2	x(x-5)(x+2)
x < -2	−	−	−	−
-2 < x < 0	−	−	+	+
0 < x < 5	+	−	+	−
x > 5	+	+	+	+

$\therefore x(x-5)(x+2) \leq 0$, when $x \leq -2$ or $0 \leq x \leq 5$.

But we must also have $-8 < x < 2$. The solution set is therefore $(-8, -2] \cup [0, 2)$. —Answer.

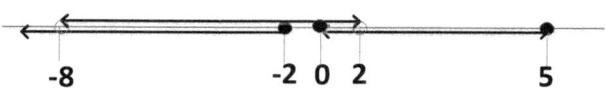

6.6 FEATURE MATHEMATICIAN

ARCHIMEDES

MATHEMATICIAN PAR EXCELLENCE

"Mathematics reveals its secrets only to those who approach it with pure love, for its own beauty."

Archimedes was born in Syracuse, Sicily in 287 BC. Most of what we know about his life comes from precious snippets slipped into the biography of the Roman soldier, Marcellus. Archimedes spent most of his life in Syracuse, but likely studied in Alexandria with Euclid's successors, whom he communicated with regularly. He is considered to be one of the greatest mathematicians of all time; ranking up top with Newton and Gauss. He is certainly the greatest mathematician of antiquity.

Archimedes was fascinated with geometry. He worked meticulously at approximating π by determining the area of circles. He constructed larger polygons outside the circle and smaller ones inside, each time increasing the number of sides of the polygons so as to get a better approximation. He used similar methods of exhaustion (slicing) to find areas under curves and volumes of solids. By making his slices thinner and thinner the approximations became exact in the limit. His use of infinitesimals anticipated modern calculus; and Renaissance mathematicians relied closely on his work. Archimedes was most proud of his discovery that the volume of a sphere is two thirds the volume of the smallest cylinder that can contain it.

Though his love was for Pure Mathematics, Archimedes was also the greatest of scientists. He was the one who, when he discovered the principle of floatation, jumped out of his bath tub and ran naked through the streets shouting, 'Eureka! I've found it!' An engineering genius, he constructed incredible war machines, which for two years defended his city against Roman attack under the command of Marcellus. Eventually, in 212 BC, the Romans took the city and Archimedes was killed by a soldier, while drawing his circles in the sand. At his request, the figures of a sphere and a cylinder were engraved on his tombstone. After his death, the golden age of Greek Mathematics soon came to an end. There were no significant new discoveries in Mathematics until the rebirth of knowledge in the 16th century and the discovery of Infinitesimal Calculus.

6.7 CHAPTER EXERCISES

1. Solve the following inequalities, expressing the solution sets in interval notation:

(i) $(2x-1)/(6x+5) < (x+1)/3x - 7, \quad x \neq 7/3, -5/6$.

(ii) $(2x-1)/(3x-4) < 1/x, \quad x \neq 0, 4/3$

2. Solve (i) $\dfrac{1}{1-x} \leq 1 + 2x, \ x \neq 1$.

(ii) $|2x-1| \geq |3x-4|$, where x is a real number.

3. Prove that the square-root of the area of a rectangle is less than or equal to one-quarter of its perimeter, with equality if and only if the rectangle is a square. Hence deduce that a square encloses the largest area among all rectangles of a given perimeter.

4. Prove that for all $a, b \in \mathfrak{R}^-$

(i) $\dfrac{1}{a} + \dfrac{1}{b} \geq \dfrac{4}{a+b}$

(ii) $(a+b)^2 \geq 4ab$

5. Prove that $\forall a, b, c \in \mathfrak{R}^+$,

$$\frac{a+b}{c} + \frac{b+c}{a} + \frac{c+a}{b} \geq 6.$$

6. (i) Prove that $\left|\dfrac{a}{b}\right| = \dfrac{|a|}{|b|}$, for all $a, b \in \mathfrak{R}$.

(ii) Prove that if a, b and c are real numbers and if $a < b$ and $b < c$, then $a < c$.

7. Prove that if a and k are fixed real numbers with $k > 0$, then

(i) $|x-a| \leq k \leftrightarrow a - k \leq x \leq a + k$.

(ii) $|x-a| \geq k \leftrightarrow x \leq a - k \text{ or } x \geq a + k$.

Illustrate these results on a number line.

CHAPTER REFERENCES

Howard Anton, *Calculus with Analytic Geometry*, 3rd ed. (New York: Wiley, 1988).

P.A. Morris, *Introduction to Algebra*, 3rd edition (Department of Mathematics, UWI, St Augustine, 1996).

Seymour Lipschutz, *Schaum's Outline of Theory and Problems of Set Theory and Related Topics, Schaum's Outline Series* (New York : McGraw-Hill, 1977).

CHAPTER SEVEN

INTEGERS, RATIONALS AND IRRATIONALS

In this chapter we extend the natural numbers to integers and study some classical topics in integer theory. These include divisibility theorems, greatest common divisors, Euclidean algorithm, prime numbers, the Fundamental Theorem of Arithmetic and Diophantine equations. We then extend the integers to the rational numbers, and investigate the existence of irrational numbers. This will conclude our introductory study of the number systems. We leave the complex numbers for next semester.

7.1 EXTENSION OF N TO Z

We saw that the set of natural numbers is the smallest inductive set. It has an identity under addition, namely zero, but the only element with additive inverse is zero. So given $a, b \in N$, the equation $x + b = a$ is not always solvable for $x \in N$. We need to expand N (imagine we did not have Z as yet).

Exercise: Define a relation $\sim$ on $N \times N$ by $(a,b) \sim (c,d) \Leftrightarrow a + d = b + c$. (In the Chapter Exercises you are asked to prove that $\sim$ is an equivalence relation on N). Then, we have:

$$[(1,1)] = \{(x,y) \in N \times N : (x,y) \sim (1,1)\}$$
$$= \{(x,y) \in N \times N : x + 1 = y + 1\}$$
$$= \{(x,y) \in N \times N : x = y\}$$
$$= \{(0,0),(1,1),(2,2),...\}.$$

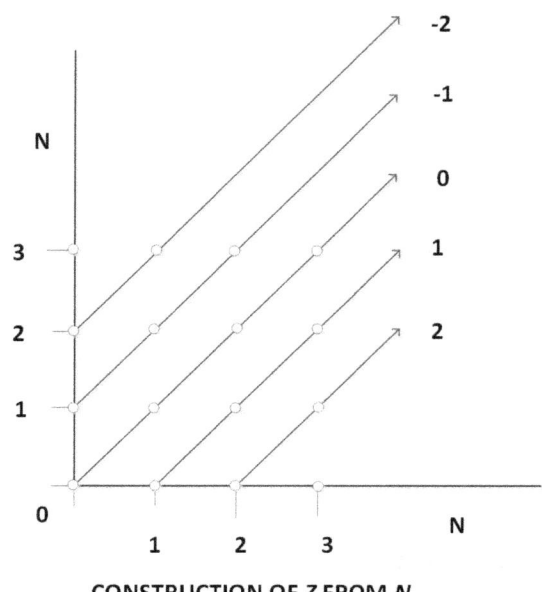

CONSTRUCTION OF Z FROM N

Similarly, $[(1,2)] = \{(x,y) \in N \times N : (x,y) \sim (1,2)\}$

$\quad = \{(x,y) \in N \times N : x+2 = y+1\}$

$\quad = \{(x,y) \in N \times N : x+1 = y\}$

$\quad = \{(0,1),(1,2),(2,3),...\}$. Call this -1.

In general, for $k \in N$, $[(1,k+1)] = \{(x,y) \in N \times N : (x,y) \sim (1,k+1)\}$

$\quad = \{(x,y) \in N \times N : x+k+1 = y+1\}$. Call this $-k$.

So $-k = [(1,k+1)] = \{(x,y) \in N \times N : x-y = -k\}$.

$\therefore -1 = [1,2]$, $-2 = [1,3]$, $-3 = [1,4]$,.... In this way we can form the negative integers.

If $k \in N$, $[(k+1,1)] = \{(x,y) \in N \times N : (k+1,1) \sim (x,y)\} = \{(x,y) \in N \times N : x-y = k\}$. Call this k. Thus, we can form the positive integers also and can regard N as a subset of Z.

We have, therefore, constructed Z from N:

$Z = \{...,[(1,3)],[(1,2)],[(1,1)],[(2,1)],[(3,1)],...\} = \{...,-2,-1,0,1,2,...\}$.

Thus, the set of equivalence classes of $N \times N$ with respect to $\sim$ is the set Z of integers. Every element of the set Z has an additive inverse.

7.2 SOME PROPERTIES OF THE INTEGERS

Definition 7.2.1: Let $a, b \in Z$ with $a \neq 0$, we say that *a divides b*, if there is an integer c such that $ac = b$. We write $a|b$. Then a is called a *divisor* or *factor* of b.

Example 1: 3 divides 6, since $6 = 3 \cdot 2$. We write $3|6$.

Note that $a|b$ is a statement. It is not the same as the division operation 'a divided by b,'—written a/b. We use the vertical bar as the symbol for 'divides'; and the slanted or horizontal bar for division.

Theorem 7.2.1: If k is a common divisor of a and b, i.e. $k|a$ and $k|b$, then $k|(xa + yb)$, for all $x, y \in Z$.

Proof: Left as an exercise.

Example 2: $7|14$ and $7|49$. $\therefore 7|(14x + 49y)$. This is true even when x, y are negative integers.

Theorem 7.2.2: For any two integers a and b, we can find integers q and $r < a$ such that $b = aq + r$ where q is called the quotient and r—the remainder when b is divided by a.

Example 3: Let $b = 14$, $a = 3$. Then $14 = 3 \cdot 4 + 2$. Hence 4 is the quotient and 2 is the remainder when 14 is divided by 3.

Definition 7.2.2: Let $a, b \in Z$. The *greatest common divisor* (G.C.D.) or *highest common factor* (H.C.F.) of a and b, written (a, b), is the largest of the common divisors of a and b.

Example 4: The common divisors of 18 and 24 are 2, 3 and 6. $\therefore (18, 24) = 6$.

Example 5: Find the G.C.D. of 4277 and 2821.

$$4277 = 1 \cdot 2821 + 1456 \quad (1)$$
$$2821 = 1 \cdot 1456 + 1365 \quad (2)$$

$$1456 = 1 \cdot 1365 + 91 \quad (3)$$
$$1365 = 15 \cdot 91 + 0 \quad (4)$$

We show below that $(4277, 2821) = 91$.

Let $d = (4277, 2821)$. Working backwards from Equation (4) and using Theorem 7.2.1, we see that 91 divides 1365 and 1456 and 2821 and 4277. Therefore, 91 is a common divisor of 4277 and 2821. This implies that $91 \leq d = \text{G.C.D.}$

To show that 91 is the greatest common divisor, note from (1) that $1456 = 4277 - 1(2821)$. Therefore, by Theorem 7.2.1, $d|1456$. Also, from Equation (2), $1365 = 2821 - 1(1456)$. Therefore, by Theorem 7.2.1, d divides 1365. Similarly, from equation (3), $d|91$. Since $d|91$, it cannot be greater than 91; i.e. $d \leq 91$. Since $d \leq 91$ and $d \geq 91$, we have $d = 91$.

So, we have an algorithm for finding $(4277, 2821)$. It is called the *Euclidean Algorithm*:

$$\begin{array}{r} 1 \\ 2821\overline{)4277} \\ \underline{2821} \end{array}$$

$$\begin{array}{r} 1 \\ 1456\overline{)2821} \\ \underline{1456} \end{array}$$

$$\begin{array}{r} 1 \\ 1365\overline{)1456} \\ \underline{1365} \end{array}$$

$$\text{G.C.D} \rightarrow \begin{array}{r} 15 \\ 91\overline{)1365} \\ \underline{1365} \\ 00 \end{array}$$

Also, from Example 5, we have:

$$91 = 1456 - 1(1365), \text{ from Equation (3)}$$
$$= 1456 - [2821 - 1 \cdot 1456], \text{ from (2)}$$
$$= 2(1456) - 2821 \quad \text{(simplifying)}$$
$$= 2(4277 - 1 \cdot 2821) - 2821, \text{ from (1)}$$
$$= 2(4277) - 3(2821) \quad \text{(simplifying)}.$$

Therefore we can write 91 as a linear combination of 4277 and 2821. This result is formalized in the following theorem.

Theorem 7.2.3 (Euclid's Theorem): If $d = (a,b)$, then $d|a$ and $d|b$; and $\exists$ integers m and n, such that $ma + nb = d$.

Example 6: From Example 5, $(4277, 2821) = 2(4277) - 3(2821)$. $\therefore m = 2$ and $n = -3$.

7.3 PRIME NUMBERS

Prime numbers are the building blocks for the set of positive integers. They have been studied extensively, for centuries.

Definition 7.3.1: Two integers are *co-prime* iff they have no common divisors other than 1.

Example 7: The numbers 8 and 9 are co-prime.

Definition 7.3.2: A positive integer, excluding 1, which has no other positive factors besides itself and 1 is called a *prime* number.

Definition 7.3.3: A *composite* number is a positive integer, excluding 1, which is not prime.

Definition 7.3.4: A prime number which divides a given integer is called a *prime divisor* of the given integer.

Example 8: The divisors of 16 are 1, 2, 4, 8 and 16. Also, 2 is the only prime divisor.

Theorem 7.3.1: If $ab = n$, where $a, b, n \in Z^+$, then $a \leq \sqrt{n}$ or $b \leq \sqrt{n}$.

Proof: Left as an exercise.

To see, for example, that 61 is a prime number, you do not have to check for divisors of all the numbers less than 61. You need only check for prime divisors up to $\sqrt{61}$. If 61 is composite, then $61 = ab$, where $a > 1$ and $b > 1$. But if $ab = 61$, (by Theorem 7.3.1) either $a \leq \sqrt{61}$ or $b \leq \sqrt{61}$, i.e. either $a \leq 7$ or $b \leq 7$. Since neither 2, 3, 5 nor 7 divide 61, no such a, b exist. $\therefore$ 61 is prime.

So we can make a list of prime numbers: 2, 3, 5, 7, 11, 13, 17, 19, 23, 29, ... Where do we stop? 2,300 years ago, Euclid showed that there is an infinitude of primes. It is the first known example of proof by contradiction:

Theorem 7.3.2: There is no greatest prime number (i.e. there is an infinitude of primes).

Proof:

Let us assume that there are finitely many primes– say, $p_1, p_2, p_3, ..., p_n$, where n is a positive integer. Consider the integer $z = p_1 p_2 p_3 ... p_n + 1$. Since $z > 1$, it must have at least one prime divisor p. If p is prime then $p = p_i$ for some $1 \leq i \leq n$. Thus p divides the product $p_1 p_2 ... p_i ... p_n$. Thus p divides the linear combination $z - p_1 p_2 p_3 ... p_n$. Thus p divides 1. But there are no primes which divide 1. Thus p is not on our finite list of prime numbers. This is a contradiction. Therefore our assumption is false. There are infinitely many primes. □

The product of the smallest n primes plus 1 is always a prime number. It is called the nth Euclid number. The study of prime numbers is intriguing. Many questions regarding them remain open. You can learn more if you take a course in Number Theory. We come now to the Fundamental Theorem of Arithmetic. The main objective is to be able to write each integer uniquely as a product of primes.

7.4 FUNDAMENTAL THEOREM OF ARITHMETIC

The Fundamental Theorem of Arithmetic states that every integer $n > 1$ can be expressed uniquely as a product of primes. It is also called the Unique Factorization Theorem. You have been using this theorem since primary school. For example, $24 = 2^3 \times 3$, $30 = 2 \times 3 \times 5$, $6413 = 11^2 \times 53$. We will leave the proof of uniqueness for Year 2 Mathematics and prove instead a simpler version of the theorem:

Theorem 7.4.1: Every integer $n > 1$ can be expressed as a product of primes.

Proof (Induction on n):

The theorem is true for $n = 2$, since $2 = 2$ (a product of primes).

Suppose it is true for $n = 3, 4, ..., k$ that these integers can be expressed as a product of primes.

Consider $n = k+1$. Either $k+1$ is a prime, in which case $k+1 = k+1$ (a product of primes), or $k+1$ is not a prime. Then $k+1 = n_1 n_2$ where $2 \leq n_1 \leq k$ and $2 \leq n_2 \leq k$ (every composite integer p can be written $p = ab$ where $a, b > 1$). But both n_1 and n_2 are products of primes by assumption, therefore $k+1$ is a product of primes. By the Principle of Mathematical Induction, the theorem is thus true for all $n > 1$. □

The Fundamental Theorem further states that for each integer $n > 1$, the prime factorization is unique. We can represent this factorization as $n = p_1 p_2 p_3 \ldots p_k$ where the p_i's $(1 \leq i \leq k)$ are not necessarily distinct or ordered (for example, $24 = 2 \times 2 \times 2 \times 3$). The prime factorization of a positive integer n (>1) can also be written:

$$n = p_1^{\alpha_1} p_2^{\alpha_2} \ldots p_r^{\alpha_r},$$

where the p_i's ($p_1 < p_2 < \ldots < p_r$) are primes and the α_i's are positive integers (for example, $24 = 2^3 \times 3$). This expression is called the prime representation of n.

We can similarly represent a rational number in this form.

Example 9: The unique prime representation of $-\dfrac{120}{169} = -\dfrac{2^3 \cdot 3 \cdot 5}{13^2} = -2^3 \cdot 3 \cdot 5 \cdot 13^{-2}$.

The prime factorization can be used, for example, to find the G.C.D. and L.C.M. of a set of integers. It is also needed to establish many theorems in number theory. You will meet it again next year. Before we extend the integers to rational numbers, we will look at one more application of the Euclidean algorithm.

7.5 LINEAR DIOPHANTINE EQUATIONS

Definition 7.5.1: A *linear Diophantine equation* is an equation of the form $ax + by = c$, where a, b, and c are integers.

If c is the greatest common divisor of a and b, then the equation is known as *Bezout's identity*. The equation has an infinite number of solutions for x and y. These can be found by applying the extended Euclidean algorithm. It follows that there are also infinitely many solutions, if c is a multiple of (a,b). If c is not a multiple of (a,b), then the equation has no integer solutions.

Example 10: The equation $4x + 6y = 3$ has no integer solutions since $(4,6)$ does not divide 3.

Example 11: Find integer solutions of $4x + 6y = 2$.

Solution: $(4, 6) = 2$. We can write 2 in the form $4a + 6b$:

$2 = -4 + 6 \Rightarrow a = -1, \ b = 1. \ \therefore (-1, 1)$ is a solution.

But $4(2) - 6(1) = 2$. Here $a = 2, \ b = -1$. Therefore $(2, -1)$ is another solution.

In fact, the equation $4x + 6y = 2$ represents a straight line in the x,y plane. The solutions are the integer points through which the line passes.

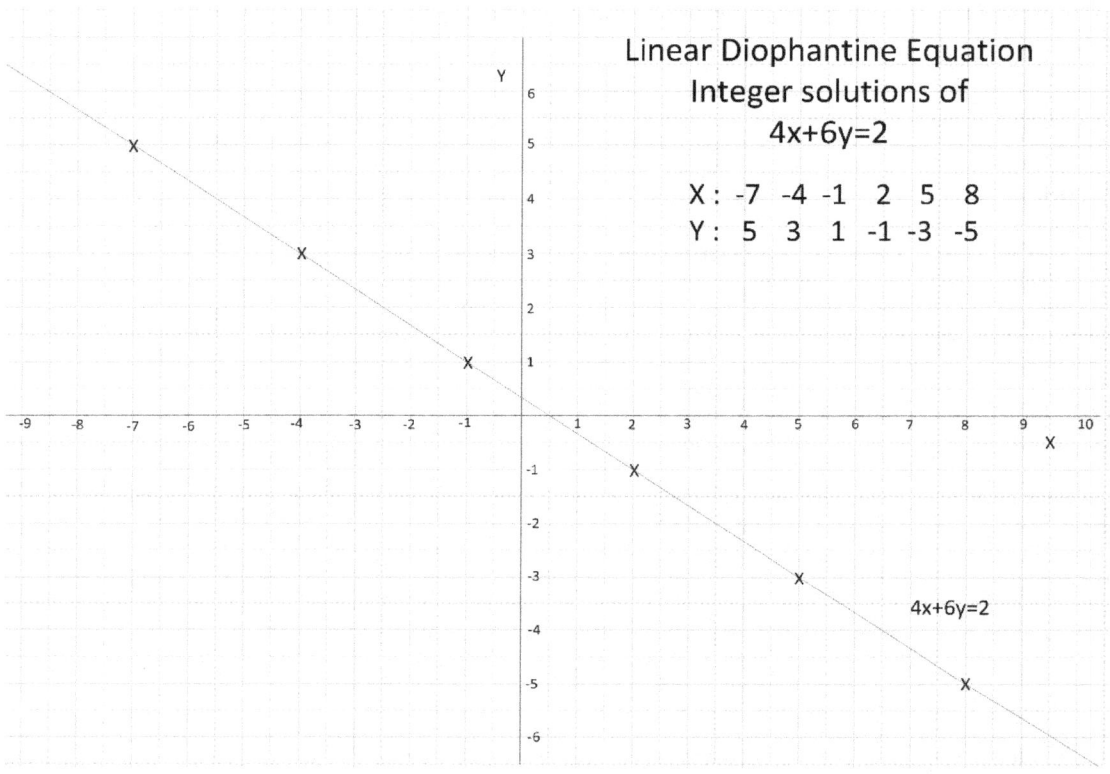

Theorem 7.5.1: If (u,v) is a solution to $ax + by = c$, where $(a,b)|c$, the equation $ax + by = c$ has the general solution

$$x = u + \frac{b}{(a,b)}z \text{ and } y = v - \frac{a}{(a,b)}z, \quad z \in Z.$$

The general solution to $4x + 6y = 2$ is therefore

$$x = -1 + \frac{6}{2}z, \quad y = 1 - \frac{4}{2}z \text{ or } (-1+3z, 1-2z), \text{ for } z \in Z.$$

We can then solve $4x + 6y = 2k$, where $k \in Z$. To solve $4x + 6y = 12$, say, since $12 = 6 \cdot 2$, there are an infinite number of integer solutions; namely, $6(-1+3z, 1-2z)$, where $z \in Z$.

Diophantine equations are named after another great Greek mathematician, Diophantus. He lived in the third century AD and he is known as the Father of Algebra. He wrote a series of books called *Arithmetica,* which contained a collection of algebraic problems that greatly influenced the development of Number Theory. He was also the first Greek mathematician to recognise fractions as numbers.

7.6 RATIONAL NUMBERS

Given $a + x = b$, where $a, b \in Z$, we can find an integer solution for x. But to solve $bx = a$, where $b \neq 0$, $x = \dfrac{a}{b}$ may not be an integer. We must extend our set of integers to make division always possible.

Define a relation on $Z \times Z$ by $(a,b) \sim (c,d)$ iff $ad = bc$. It can be shown that $\sim$ is an equivalence relation (see question 8 of the Chapter Exercises). Then:

$$[(1,2)] = \{(x,y) \in Z \times Z : (x,y) \sim (1,2)\} = \{(x,y) : 2 \cdot x = 1 \cdot y\}$$

$$= \left\{(x,y) \in Z \times Z : \frac{x}{y} = \frac{1}{2}\right\} = \{...,(-1,-2),(0,0),(1,2),(2,4),(3,6),...\}.$$

Let $\frac{1}{2}$ denote $[(1,2)] = [(2,4)] = [(3,6)]$ and, in general, let $\frac{a}{b}$ denote $[(a,b)]$, $b \neq 0$.

The set of equivalence classes of $Z \times Z$ gives the set of rational numbers $Q = \left\{\dfrac{a}{b} : a, b \in Z, b \neq 0\right\}$. This coincides with the definition of Q given in Chapter 4.

Theorem 7.6.1: Between any two rational numbers there is another rational number.

Example 12: Between $\frac{1}{2} = \frac{6}{12}$ and $\frac{1}{3} = \frac{4}{12}$, we clearly have the rational number $\frac{5}{12}$.

i.e. $\frac{1}{3} \leq \frac{5}{12} \leq \frac{1}{2}$.

Clearly, there is a vast array of rational numbers. But can we count them? Can we even count the positive rational numbers, like we can count $Z^+ = \{1, 2, 3,...\}$? If we put 1 first, which rational number would we put next? Below is one attempt.

Let us set out the positive rational numbers as fractions in a table like this:

Q^+	DENOMINATOR						
N U M E R A T O R	$\frac{1}{1}$ (1)	$\frac{1}{2}$ (2)	$\frac{1}{3}$ (4)	$\frac{1}{4}$ (7)	$\frac{1}{5}$ (11)	$\frac{1}{6}$	...
	$\frac{2}{1}$ (3)	$\frac{2}{2}$ (5)	$\frac{2}{3}$ (8)	$\frac{2}{4}$ (12)	$\frac{2}{5}$	$\frac{2}{6}$	...
	$\frac{3}{1}$ (6)	$\frac{3}{2}$ (9)	$\frac{3}{3}$ (13)	$\frac{3}{4}$	$\frac{3}{5}$	$\frac{3}{6}$	...
	$\frac{4}{1}$ (10)	$\frac{4}{2}$ (14)	$\frac{4}{3}$	$\frac{4}{4}$	$\frac{4}{5}$	$\frac{4}{6}$	...
	$\frac{5}{1}$ (15)	$\frac{5}{2}$	$\frac{5}{3}$	$\frac{5}{4}$	$\frac{5}{5}$	$\frac{5}{6}$	
	⋮	⋮	⋮	⋮	⋮	⋮	⋮

Each positive rational is in some cell of the table. We can count them as indicated by the bracketed numbers (1), (2), (3), ... The rational numbers are countable! This means they can be put in a 1-1 correspondence with the set of integers—more of this next year!

Decimal Representation

Rational numbers have repeating decimal representation. for example, $\frac{1}{3} = .33\dot{3}$. We obtain the decimal representation by dividing the numerator by the denominator. Given the decimal representation, how do we find the fraction? We learnt this in primary school. For example, $.125 = 125/1000$. For recurring decimals, the method is less known. The box, at right, illustrates how to change $.33\dot{3}$ into a fraction.

$$10x = 3.333...$$
$$x = .333...$$
$$\overline{9x = 3.000}$$
$$\therefore x = \frac{3}{9} = \frac{1}{3}$$

7.7 IRRATIONAL NUMBERS

We can now solve linear equations of the form $bx = a$, where $b \neq 0$. But, what about the equation $x^2 = 2$ in Q?

Theorem 7.7.1: There is no rational number x satisfying $x^2 = 2$.

Proof (by contradiction):

Suppose that $x = \dfrac{m}{n}$ ($m, n \in Z, n \neq 0$) satisfies $x^2 = 2$, where we take $\dfrac{m}{n}$ in its lowest form (i.e. m and n have no common factors). Now $\dfrac{m^2}{n^2} = x^2 = 2$. $\therefore m^2 = 2n^2.$ $\Rightarrow m^2$ is even. $\Rightarrow m$ is even. So let $m = 2k$, where $k \in Z$. Then $m^2 = 4k^2$. $\therefore 4k^2 = 2n^2$. $\therefore n^2 = 2k^2$. $\Rightarrow n^2$ is even; so n is even. $\therefore m$ and n are even. This contradicts m and n having no common factors. $\therefore x$ cannot be written in form a/b. □

So $\sqrt{2}$ exists; but it is not a rational number. We say it is irrational. Irrational numbers were first discovered in the 5th century BC, by a Pythagorean, named Hippasus. It caused quite a crisis among the Pythagoreans. Hippasus was thrown overboard at sea for his unwelcomed discovery.

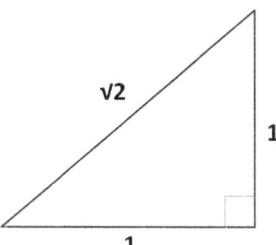

Apart from $\sqrt{2}$, there are many more irrational numbers. In fact, if p is any prime number, then $\sqrt{p}$ is irrational. We can also add any rational number to an irrational number and get another irrational number (e.g. $1 + \sqrt{2}$ is irrational). Or, take a number like 0.101001000100001000001.... It can go on and on without ever repeating. It is irrational. Then there are those incredible numbers, such as, π and e.

There are, in fact, more irrational numbers than there are rational numbers. If we tried to count the irrational numbers we would run out of integers to count them! This makes the set of real numbers uncountable. In 1873 Georg Cantor, the inventor of Set Theory, proved it.

The irrational numbers can all be represented by points on the number line but we can never measure them exactly. Here is a way to approximate $\sqrt{2}$. The method works for any prime.

Example 13: An approximation for $\sqrt{2}$:

$\sqrt{2}$ is a root of the equation $x^2 - 1 = 1$ or $(x-1)(1+x) = 1$. Therefore

$$x = 1 + \frac{1}{1+x} \qquad (1)$$

Substituting $1 + \frac{1}{1+x}$ successively for x on the right hand side of equation (1) we get:

$$x = 1 + \cfrac{1}{1 + 1 + \cfrac{1}{1+x}}$$

$$x = 1 + \cfrac{1}{1 + 1 + \cfrac{1}{1 + 1 + \cfrac{1}{1+x}}}$$

$$= 1 + \cfrac{1}{1 + 1 + \cfrac{1}{1 + 1 + \cfrac{1}{1 + 1 + \cfrac{1}{1+x}}}}$$

So $x = 1 + \cfrac{1}{2 + \cfrac{1}{2 + \cfrac{1}{2 + \cfrac{1}{2...}}}}$. We can continue the fraction to the desired degree of accuracy.

An approximation for $\sqrt{2}$ is $1 + \cfrac{1}{2 + \cfrac{1}{2 + \cfrac{1}{2.5}}} = 1 + \cfrac{1}{2 + \cfrac{1}{2.4}} = 1 + \cfrac{1}{\frac{5.8}{2.4}} = 1 + \frac{2.4}{5.8} = 1.41$.

7.8 FEATURE MATHEMATICIAN

EUCLID OF ALEXANDRIA

FATHER OF GEOMETRY

"Mathematics is the queen of the Sciences and Number Theory is the queen of Mathematics."

A painting of Euclid by Justus van Gent, 1474.

Euclid lived in Alexandria, Egypt during the reign of Ptolemy 1 (around 300BC). Very little is known of his life. Most likely, he studied at Plato's academy in Athens; then taught at Ptolemy's university for 20-30 years, while writing his 13 volume *Elements* and other works. *Elements* gives an exhaustive account of Mathematics at that time. It is best known for its geometrical results, including the works of Pythagoras, the geometry of lines and planes, areas of regular plane figures and circles. It also includes Number Theory and Arithmetic, with proofs of many of the theorems presented in this chapter, as well as Euclid's own work on irrational numbers.

Although many of his results originated with earlier mathematicians, one of Euclid's accomplishments was to present them in a single logically coherent work. These have been preserved and handed down to us. Euclid's *Elements* served as the main textbook for the teaching of Mathematics for 2000 years! His system of definition, axiom, theorem, and proof, remains the basic format for presenting Mathematics today.

As we have seen in this chapter, Euclid did a lot of work in Number Theory; but he did even more in Geometry. This earned him the title 'Father of Geometry.' He constructed, for example, a regular 12-sided dodecahedron, starting with a cube. The elegant proofs in Geometry, which up until a few decades ago, were taught at high school, are Euclid's. Our front cover fresco shows Euclid at lower right, drawing a geometrical diagram for his students.

Euclid valued learning. When one of his students asked him "what shall I get from learning all these things?" he called his servant and said "give him three pence, since he must make a gain of what he learns."[1] The story is also told that when the king asked him if there was a shorter way to study Geometry, he answered, 'There is no royal road to Geometry." The study of Mathematics is hard work; but nothing worth having, comes easy.

Euclid was followed by Archimedes; and centuries later, by Diophantus. Then came the Dark Ages and the 1000 year decline of learning. But the foundations of Mathematics had been firmly laid, rebirth and explosion of knowledge were to come. We have already seen the rise of brilliant mathematicians like Euler and Descartes. Next semester, we will meet more pioneers of this age.

[1] James R. Newman, *The World of Mathematics, Volume 1,* Novelio & Co. Ltd., London, 1956.

7.9 CHAPTER EXERCISES

1. A relation $\sim$ is defined on the set $N \times N$, by $(a,b) \sim (c,d) \Leftrightarrow a+d = b+c$.

(i) Show that $\sim$ is an equivalence relation.

(ii) Find (a) $[(1,1)]$, (b) $[(1,2)]$, (c) $[(1, k+1)]$, $k \in N$, (d) $[(k+1,1)]$.

2. (i) Use the Euclidean Algorithm to find the G.C.D. of 490 and 1113.

(ii) Find m and n such that $490m + 1113n =$ G.C.D.

3. Use Euclid's Algorithm to express $\dfrac{13}{120}$ in the form $\dfrac{p}{8} + \dfrac{q}{15}$

[Hint: Find the G.C.D. of 8 and 15 and express it in the form $ma + nb$].

4. Use the Euclidean Algorithm to find the integer solutions of $7x + 9y = 5$.

5. Give the prime representation of (i) 2728 and (ii) $\dfrac{330}{169}$.

6. (i) Prove by contradiction, that if $ab = n$, $a, b, n \in Z^+$, then either $a \leq \sqrt{n}$ or $b \leq \sqrt{n}$.

(ii) Show that the last digit of an odd prime number (except 5) must either be 1, 3, 7, or 9.

7(i) Prove that if k is a common divisor of a and b, then $k | (xa + yb)$, $\forall\, x, y \in Z$.

(ii) Prove that if $d = (a,b)$, then an integer k is a common divisor of a and b, iff $k | d$.

(iii) Prove that if a, b and c are integers such that $a | b$ and $b | c$, then $a | c$.

(iv) Prove that if a, b and c are positive integers such that $(a,b) = 1$ and $a \mid bc$, then $a \mid c$.

8. Define a relation $\sim$ on $Z \times Z$ by

$$(a,b) \sim (c,d) \text{ iff } ad = bc.$$

(i) Show that $\sim$ is an equivalence relation (ii) Find $[(a,b)]$ where $b \neq 0$.

9. Prove that if a^2 is an even integer, then a is even.

10. Prove that $\sqrt{3}$ is irrational, then use the method of continued fractions to approximate it to 2 decimal places.

CHAPTER REFERENCES

Herbert B. Enderton, *A Mathematical Introduction to Logic*, 2 ed. (New York: Academic/Elsevier, 2006).

P.A. Morris, *Introduction to Algebra*, 3rd edition (Department of Mathematics, UWI, St Augustine, 1996).

Wissam Raji, *An Introductory Course in Elementary Number Theory* (Lulu.com, 2012).

APPENDIX 1

COURSE DESCRIPTION

In this appendix we provide the course description for the MATH 1152—Sets and Number Systems Course of the Department of Mathematics and Statistics, The University of the West Indies, St Augustine, Trinidad and Tobago. A more detailed course outline is available from the department.

SETS AND NUMBER SYSTEMS–COURSE OUTLINE

Course Rationale:

Mathematics is a powerful tool for solving practical problems and is a highly creative field of study, combining logic and precision with intuition and imagination. The ability to employ mathematical reasoning is a fundamental skill for any well-educated individual in the pure and applied sciences. A basic knowledge of mathematics is needed to provide the necessary framework for solving problems in fields such as medicine, management, economics, computer science, physics, psychology, engineering, and the social sciences.

MATH 1152—Sets and Number Systems is an introductory level course designed to introduce science students to the basic concepts of Algebra and to provide a solid foundation for those interested in further studies in Mathematics. It is a prerequisite for students taking advanced level courses in Mathematics, Statistics, or Actuarial Science, as well as, for those majoring in Physics and other Mathematics-related Sciences.

Prerequisites:

A necessary prerequisite for the course is two units of (CAPE) Mathematics or any of its equivalents (such as, GCE Advanced Level Mathematics).

Credits:

This is a semester long course with 2 teaching hours per week plus 1 hour of tutorial. The student earns 3 credits of Level 1 Mathematics upon successful completion of the course.

Some Specific Learning Outcomes:

Upon successful completion of the course, students will be able to:

- Analyse arguments via the rules of logic.
- Formulate mathematical proofs by using the rules of logic.
- Prove that two sets are equal.
- Use Venn Diagrams to test the validity of arguments.
- Test for the basic properties of equivalence relations, functions and binary operations.
- State the field axioms as basic properties of real numbers.
- Test for some basic properties of real numbers.
- Prove results using the principle of mathematical induction.
- Prove basic theorems on absolute value.
- Solve inequalities involving quotients and absolute values.
- Apply the Arithmetic Mean-Geometric Mean Inequality.
- Discuss basic properties of natural numbers, integers, rationals, and real numbers.
- Use the Euclidean Algorithm to find the greatest common divisor of two or more integers.
- Prove simple theorems about prime numbers and divisibility of integers.
- Solve linear Diophantine equations.
- Prove that $\sqrt{2}$ cannot be rational.

Course Content:

Logic:
- Statements in Mathematics. Manipulation of statements: negation, conjunction, disjunction, implication and double implication.
- Truth tables, logical equivalence, converse and contrapositive, tautology and contradiction, quantifiers.

- Classical arguments and proof of validity.
- Proof by contradiction, mathematical proofs.

Set Theory:
- The notion of a set as the fundamental structure of Mathematics. Equality of sets. Subsets, unions, intersections, complement and set difference. Formal proofs. Venn Diagrams.
- Illustration of validity of arguments using sets. Set Algebra and De Morgan's Laws.

Functions:
- The Cartesian product of sets.
- Definition of a function. Proof of injectivity and surjectivity of functions. Composition and inverse of a function.

Relations:
- A relation as a generalization of a function. Properties of relations. Reflexive, symmetric and transitive relations. Equivalence relations and partitions of sets.

Binary Operations:
- Properties of binary operations. Commutativity, associativity, distributivity, closure, existence of identity and inverse elements.

The Natural Numbers:
- Definition of Natural number.
- The Principle of Mathematical Induction.

The Integers:
- Extension of Natural numbers to Integers. Basic facts regarding divisibility. The greatest common divisor and the Euclidean Algorithm.
- The infinitude of the primes. The Fundamental Theorem of Arithmetic. Linear Diophantine Equations.

The Rational Numbers:
- Extension of Integers to Rationals. The proof that $\sqrt{2}$ cannot be rational.

The Real Numbers:
- The Field Axioms.
- Solution of linear and non-linear inequalities in R. The absolute value. The Triangle Inequality. The Arithmetic Mean-Geometric Mean Inequality.

APPENDIX 2

PAST EXAMINATION PAPERS

In this section we present the Math 1152 – Sets and Number Systems Examinations of the Department of Mathematics, The University of The West Indies, St. Augustine. It will be good to practice these questions even if you are not sitting the exam. The time allotted is two hours. Full marks may be obtained for correct answers to four questions. Marks per question are shown in square brackets. The use of non-programmable calculators is permitted.

EXAMINATIONS OF DECEMBER 2012–MATH 1152

SECTION A

1 (a) (i) Give the meaning of the term "statement."

Let A and B be mathematical statements. Show that:

(ii) $A \Rightarrow B$ is not logically equivalent to $B \Rightarrow A$.

(iii) $(A \Rightarrow B) \Leftrightarrow (\sim B \Rightarrow \sim A)$ is a tautology.

(iv) $\sim (A \Rightarrow B) \equiv A \wedge \sim B$. [4]

(b) Give the converse, the contrapositive and the negation of the statement "All men are smart."

In each case show how you derived your answer.

Which of these is equivalent to the given statement? [6]

(c) First convert the following argument to the language of logic or set theory and then test the validity. [5]

All men are smart
Monkeys are smart
_____
All men are monkeys

2 (a) Let A and B be sets in a given universe. Prove formally that $A - B = A \cap B'$.

Illustrate with Venn Diagrams. [4]

(b) Use set algebra to prove that $A - (A \cap B) = (A \cup B) - B$.

State any theorems that are used. [4]

(c) Let $f : R^+ \cup \{0\} \to R^+ \cup \{0\}$ be defined by $f(x) = x^2$.

Prove that f is bijective and determine $f^{-1}(x)$. [7]

SECTION B

3 (a) Let $*$ be a binary operation defined on the set R of real numbers by

$$x * y = x + y + xy$$

(i) Determine if $*$ is commutative.

(ii) Determine if $*$ is associative.

(iii) Determine if there is an identity element with respect to $*$.

(iv) Determine which elements of R have an inverse under $*$. [8]

(b) R is a relation defined on Z as follows:

$$\forall\, m, n \in Z, \quad m\,R\,n \iff m^2 - n^2 \text{ is even}$$

Prove that R is an equivalence relation and find $[1]$ and $[2]$. [7]

4 (a) Solve $\dfrac{x-3}{\left|x^2 - 9\right|} < 1$, for $x \in R$ and $x \neq \pm 3$. [6]

(b) Let x and y be real numbers with $x \geq 0$. Prove that $x + y^2 \geq 2\,y\,\sqrt{x}$. [1]

(c) Prove that $|a+b| \leq |a|+|b| \quad \forall\ a,b \in R$. [2]

(d) Prove by the use of mathematical induction that $\sum_{k=0}^{n} \alpha^k = \dfrac{1-\alpha^{n+1}}{1-\alpha}$, $\forall \alpha \neq 1$ and $n \in N$. [6]

SECTION C

5 (a) State Euclid's Theorem. Hence find all solutions to the equation $310\,x + 510\,y = 30$, for $x,y \in Z$. [6]

(b) Prove that $\sqrt{2}$ is not a rational number. [4]

(c) State the Fundamental Theorem of Arithmetic. Use this theorem to prove that if x^2 divides y^2 then x divides y, where x and y are integers. [5]

6 (a) Let $z = x + i\,y$ be a complex number. If $\dfrac{z-i}{z+1}$ is purely real, find an equation in x and y that describes the path traced out by z in the Argand diagram. [3]

(b) Express $\sqrt{3}+i$ and $1-i$ in the (r,θ) form and hence find the modulus and principal argument of $z = \dfrac{(\sqrt{3}+i)^{17}}{(1-i)^{10}}$. [6]

(c) Let z and w be complex numbers. If $|z|=1$, show that $\left|\dfrac{z-w}{1-w\bar{z}}\right|=1$. [3]

(d) Find the cube roots of 1 **without** using De Moivre's theorem. [3]

END OF PAPER

EXAMINATIONS OF DECEMBER 2013–MATH 1152

SECTION A

1 (a) Use truth tables to determine which of the following statements are logically equivalent:

(i) If I go you will go.

(ii) If you go I will go.

(iii) If you do not go I will not go.

(iv) I am going but you are not going.

(v) Either you go or I don't go. [10]

(b) Write the negation of the statement

"Some truths are absolute"

If the given statement is true what can we say about its negation? [5]

(c) Determine the validity of the following argument: [5]

P_1 : If I speak, I am condemned.

P_2 : If I stay silent, I am damned.

C : If I am not condemned, I am damned.

(d) Suppose we know that x is not less than y. Prove, by contradiction, that $x+1$ is not less than $y+1$. [5]

2 (a) (i) Let A, B and C be sets in a given universe. Prove formally that if A is a subset of B and B is a subset of C, then A is a subset of C. [6]

(ii) Let D, E and F be sets in a given universe.

Prove that if $D \cap E \subseteq F'$ and $D \cup F \subseteq E$ then $D \cap F = \emptyset$. [6]

(b) Use set algebra to prove that $(A \cup B)^c \cup (A-B) = B^c$ ('c' means complement). (Please state all laws that are used). [7]

(c) Convert into the language of sets and use Venn Diagrams to determine whether or not the following argument is valid. [6]

p_1 : All politicians love power.

p_2: Some men who love power love money.

C : Some politicians love money.

SECTION B

3 (a) Let S be the subset of real numbers given by $S = \{x \in R : x \geq 1\}$. Let $*$ be an operation on S, defined by

$$x * y = 1 + \sqrt{(x-1)^2 + (y-1)^2}$$

Prove that:

(i) S is closed under $*$. [2]

(ii) $*$ is commutative. [2]

(iii) $*$ is associative. [4]

(iv) S has an identity element for $*$ and determine which (if any) elements of S have inverses under $*$. [5]

(b) Let Z be the set of integers. A relation $\sim$ is defined on $Z \times Z$ as follows:

$$(a, b) \sim (c, d) \text{ if and only if } a^2 + d^2 = b^2 + c^2$$

(i) Determine, with proof, if $\sim$ is an equivalence relation on $Z \times Z$. [2, 2, 4]

(ii) Find the equivalence class of the element $(1, 2)$, if it is possible to do so. [4]

4 (a) Prove the following equality by the Principal of Mathematical Induction:

$$1.2 + 2.3 + 3.4 + 4.5 + 5.6 + 6.7 + ,..., + n(n+1) = \frac{n(n+1)(n+2)}{3}$$

for all positive integers n. [6]

(b) f and g are two functions defined on the set of real numbers R as

$$f(x) = \begin{cases} x^2 \ ; \ x \geq 0 \\ x - 1 \ ; \ x < 0 \end{cases} \text{ and } g(x) = x + 1.$$

Find the composition of f and g (i.e. $(f \circ g)(x)$), with its domain. [3]

(c) Let R be the set of real numbers with A and B being subsets of R. Let f be a relation from A to B defined by

$$f(x) = \frac{x-3}{2x+1}$$

(i) Is f a function on R? Give reasons for your answer. [2]

(ii) Give the largest set A for which f is a function. [2]

(iii) In case (ii), determine if f is $1-1$. [4]

(iv) In case (ii), give the largest set B, such that f is both $1-1$ and onto. [5]

(v) In case (iv), give a formula that defines the inverse of f. [3]

SECTION C

5 (a) Solve the inequality $x^2 - 3|x-1| > 1$, for real values of x. [9]

(b) Let x and y be positive real numbers. Prove that [9]

$$\frac{x^2}{y^2} + \frac{y^2}{x^2} + 6 \geq \frac{4x}{y} + \frac{4y}{x}$$

(c) Let a and b be real numbers. Prove that $|a| - |b| \leq |a - b|$. [7]

6(a) Let a, b and c be integers. Prove that if a divides b and b divides c, then a divides c. [5]

(b) Find all integer solutions in the Diophantine equation $21x + 14y = 147$. [9]

(c) Let n be a positive prime integer. Prove that $\sqrt{n}$ is irrational. [6]

(d) Let m and n be integers such that $m|n$ and $n|m$. Prove that $|m| = |n|$. [5]

END OF PAPER

EXAMINATIONS OF DECEMBER 2014–MATH 1152

In This Paper Full Marks May Be Obtained For Correct Answers To All Five Questions.

1. (a) Let A be the following statement: "If it is big then it must be heavy."

    (i) Write (in words) the converse of A. [2]

    (ii) Write (in words) the contrapositive of A. [2]

    (iii) Write a negation of A. [2]

    (iv) Rewrite A using the word "necessary." [2]

(b) Let p and q be propositions. Use a truth table to determine whether $q \Rightarrow (p \Rightarrow q)$ is logically equivalent to $p \Rightarrow (p \Rightarrow q)$. [3]

(c) Let p, q, and r be propositions.

Determine the validity of the following argument: [4]

1. $q \Rightarrow \sim r$
2. $\sim p \Rightarrow \sim r$
3. q
   ___
   C. $\sim r$

(d) Suppose that x is a real number.

Prove by contradiction that if $x^3 + x = 0$, then $x = 0$. [5]

2. (a) Let A and B be sets in a given universe.

(i) Prove formally that $(A \cap B)^c \subseteq A^c \cup B^c$. ("c" denotes complement) [4]

State a further condition necessary to show that $(A \cap B)^c = A^c \cup B^c$. [1]

(ii) Prove formally that $A \cup \emptyset \subseteq A$. [3]

(b) Prove by the use of **set algebra** that

$$(A \cup B) \cap (A^c \cup B^c) = (A - B) \cup (B - A)$$

Please state any theorems used. [6]

(c)(i) What is meant by saying that an argument is valid? [1]

(ii) Convert the following argument into the language of set theory and use Venn Diagram(s) to test its validity: [5]

P_1: All men are logical.
P_2: Socrates is logical.
___
Conclusion: Socrates is a man

3. (a) A binary operation $*$ is defined on the set Z of integers as follows:

$$a * b = a + b - 2$$

Prove that

(i) * is commutative

(ii) * is associative

(iii) Z has an identity element with respect to *

(iv) Every element of Z has an inverse under * [8]

(b) (i) Let R be a relation to a set A. Define the following terms:

1. Reflexive relation

2. Symmetric relation

3. Transitive relation [3]

(ii) Let Z be the set of all integers. Define a relation R on Z by

$$a \; R \; b \text{ if and only if } a+b \text{ is divisible by } 2$$

Show that R is an equivalence relation and determine the equivalence classes of the elements 0 and 1. [9]

4. (a) Prove the following inequality by the Principle of Mathematical Induction. [9]

$$\frac{1}{1.2} + \frac{1}{2.3} + \frac{1}{3.4} + \frac{1}{4.5} + ,..., + \frac{1}{n(n+1)} = \frac{n}{n+1}; \quad n > 0$$

(b) Let R^+ be the set of all positive real numbers and f be a function defined as:

$$f : R^+ - \{1\} \to R^+ \text{ and } f(x) = \frac{x^2}{1+x^2}$$

Determine whether or not:

(i) f is 1-1 [4]

(ii) f is onto [6]

(iii) If f is one to one, then define the inverse function $f^{-1}(x)$ and state its domain. [1]

5. (a) Solve the inequality $\dfrac{|x-1|}{1+x} > 0$. [7]

(b) Let x and y be positive real numbers. Prove that $x^3 + y^3 \geq x^2 y + xy^2$. [6]

(c) Let d be the greatest common divisor of the integers a and b.

Prove that an integer k is a common divisor of a and b if and only if k divides d. [7]

END OF PAPER

EXAMINATIONS OF DECEMBER 2015–MATH 1152

1. (a) Let p and q be propositions.

Determine whether or not $(q \Rightarrow p) \Rightarrow (p \vee q)$ is a tautology. [6]

(b) Let A be the statement: "Only if Paul goes to school he will get his books and allowance money."

(i). Write the above using the words "if" and "then." [2]

(ii). Write the converse of the contrapositive of A using the word "sufficient." [4]

(iii). Write the negation of the following statement: [2]

"If Bob has two phones then Maria will get a new laptop for Christmas."

(c). Let p, q, r and s be propositions. Determine the validity of the following argument: [6]

$$P_1 : p \vee s$$
$$P_2 : s \wedge r$$
$$P_3 : p \Rightarrow \sim q$$
$$P_4 : q \Rightarrow \sim r$$
$$\overline{Con : r \Rightarrow p}$$

2. (a) (i) Let A, B and C be sets in a given universe.

By giving a counter-example, show that $A \cap (B \cup C) \neq (A \cap B) \cup C$. [3]

(ii). Draw Venn diagrams to depict $A \cap (B \cup C)$ and $(A \cap B) \cup C$. (Use separate diagrams).

Do your Venn diagrams suggest that $A \cap (B \cup C) \subseteq (A \cap B) \cup C$ or $(A \cap B) \cup C \subseteq A \cap (B \cup C)$? (No proof is needed.) [4]

(b). Use Set Algebra to show

(i) $(A \cap B) \cup \{(A \cap C) \cup (A^c \cap C)\} = (A \cap B) \cup C$, where A^c denotes the complement of A. [4]

(ii). $(A \cap B) \cup \{(A \cap C) \cup (A^c \cap C)\} = \{A \cap (B \cup C)\} \cup (A^c \cap C)$ [4]

(c). Prove formally that

$$(A \cap B)^c \subseteq A^c \cup B^c \quad (^c \text{ denotes complement}).$$ [5]

3. (a). Let $P(X)$ be the set of all subsets of a set X and let $*$ be the operation of intersection defined on $P(X)$, i.e. for all elements A and B in $P(X)$ we have

$$A * B = A \cap B$$

Answer the following questions by giving reasons:

(i) Is $P(X)$ closed under $*$? [1]

(ii) Is $*$ commutative? [1]

(iii) Is $*$ associative? [2]

(iv) Is there an identity element? [2]

(v) Are there any elements of $P(X)$ with inverses? [3]

(b) Let $A = \{a, b, c\}$ and $R = \{(a,b), (b,c), (a,c), (c,c)\}$

Answer the following questions with reasons:

(i) Is R a relation on A? [1]

(ii) Find R^{-1}, the inverse of R. [1]

(iii) Is R a function of A? [2]

(iv) Extend R to an equivalence relation on A, by adding the least possible number of new elements and identify the equivalence classes. [3]

(v) Find a relation on A that is reflexive, symmetric and not transitive. [2]

(vi) Find a relation on A that is reflexive and neither symmetric nor transitive. [2]

4. (a) Use Mathematical induction to prove the following equality: [9]

$$1^2 - 2^2 + 3^2 - 4^2 + \ldots + (-1)^{n+1} n^2 = (-1)^{n+1} \frac{n(n+1)}{2}$$

(b) Let R be the set of real numbers. Let $f : R - \{5\} \to R$ be defined by:

$$f(x) = \frac{3x}{x-5}$$

(i) Determine whether f is 1-1. [3]

(ii) Determine whether f is onto. [3]

(iii) Find the range of f. [1]

(iv) Does f^{-1} exist? Give reasons for your answer. [2]

(v) If f^{-1} exists, then define f^{-1}. [2]

5. (a) Let x and y be positive real numbers. Prove that if $x > y$ then [6]

$$x^3 + 3xy^2 > 3x^2 + y^3$$

(b) Solve the inequality $2x < |x-1|$, for real numbers x. [8]

(c) Let $x \equiv y \pmod{m}$, where x and y are integers. This reads "x is congruent to y modulo m" and means that $x - y$ is divisible by m.
Show that $ax \equiv ay \pmod{m}$ for any integer a. [6]

END OF PAPER

APPENDIX 3

SOLUTIONS TO PAST EXAM PAPERS

In this section we present the solutions to the Math 1152 – Sets and Number Systems Examinations of Department of Mathematics, The University of The West Indies, St. Augustine, for the years 2012-2015. You should not look at these solutions until you have honestly attempted the questions and want to check if your answers are correct. It is bad practice to try to 'swot' or memorize the solutions. The questions test your understanding of the subject matter and not your ability to reproduce answers.

SOLUTIONS TO DECEMBER 2012–MATH 1152

SECTION A

1. (a) (i) A statement is a sentence which can either be true or false but not both.

(ii) $A \to B \not\equiv B \to A$

A	B	$A \to B$	$B \to A$
T	T	T	T
T	F	F	T
F	T	T	F
F	F	T	T

Since $A \to B$ and $B \to A$ have different truth values for the same truth values of their simple statements, they are not logically equivalent.

(iii)

A	B	$A \to B$	$\sim B$	$\sim A$	$\sim B \to \sim A$	$A \to B \Leftrightarrow \sim B \to \sim A$
T	T	T	F	F	T	T
T	F	F	T	F	F	T
F	T	T	F	T	T	T
F	F	T	T	T	F	T

Since $A \to B \Leftrightarrow \sim B \to \sim A$ is always true, it is a tautology.

(iv)

A	B	$A \to B$	$\sim (A \to B)$	$\sim B$	$A \wedge \sim B$
T	T	T	F	F	F
T	F	F	T	T	T
F	T	T	F	F	F
F	F	T	F	T	F

Since $\sim (A \to B)$ and $A \wedge \sim B$ always have the same truth value, they are logically equivalent.

(b)   "All men are smart"

Let p : it is a man , q : it is smart . The given statement is $p \to q$.

(i) Converse $q \to p$:

'It is smart $\to$ it is a man', or 'all smart creatures are men', or 'it is smart only if it is a man.'

(ii) Contrapositive $\sim q \to \sim p$:

'If it is not smart, it is not a man' or 'stupid creatures are not men.'

(iii) $\sim (p \to q) \equiv p \wedge \sim q$:

'It is a man and it is not smart' or 'some men are stupid.'

Alternative Solution: Let $p(x)$: 'man x is smart'. The given statement is $\forall x, p(x)$.

Negation: $\sim(\forall x, p(x))$ or $\exists x (\sim p(x))$, i.e. 'There is a man who is not smart.'

The contrapositive is equivalent to the given statement.

(c).

$$\begin{array}{c} \text{All men are smart} \\ \text{Monkeys are smart} \\ \hline \text{All men are monkeys} \end{array}$$

Let p : it is a man, q : it is smart, r : it is a monkey.

In the language of logic, the given argument is: $\begin{array}{c} p \to q \\ r \to q \\ \hline p \to r \end{array}$

We can determine its validity using truth tables.

Or, in the language of sets, let $P = \{\text{men}\}$, $Q = \{\text{smart creatures}\}$, $R = \{\text{monkeys}\}$.

The given argument is: $\begin{array}{c} P \subseteq Q \\ R \subseteq Q \\ \hline P \subseteq R \end{array}$

Using Venn diagrams:

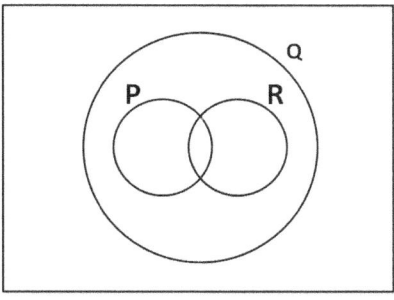

P is not a subset of R

There is a case when the premises are true but the conclusion is false. ∴ argument is invalid.

2. (a) $A - B = A \cap B^c$

Proof:

(i) Let $x \in A - B$.

$x \in (A - B) \Rightarrow x \in A$ and $x \notin B$

$\Rightarrow x \in A$ and $x \in B^c$

$\Rightarrow x \in (A \cap B^c)$. $\therefore A - B \subseteq A \cap B^c$.

(ii) Let $x \in (A \cap B^c)$.

$x \in (A \cap B^c) \Rightarrow x \in A$ and $x \in B^c$

$\Rightarrow x \in A$ and $x \notin B$

$\Rightarrow x \in (A - B)$. $\therefore A \cap B^c \subseteq A - B$.

From (i) and (ii) we $A - B = A \cap B^c$. □

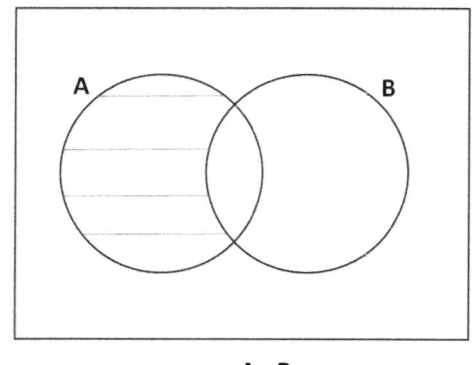

A - B

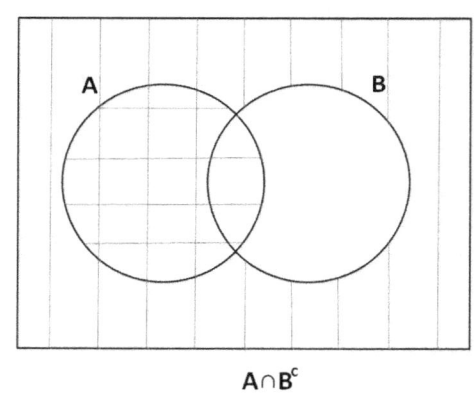

A∩B^c

(b) $A - (A \cap B) = (A \cup B) - B$

Proof:

LHS $= A - (A \cap B) = A \cap (A \cap B)^c$ (*Theorem*: $X - Y = X \cap Y^c$)

$= A \cap (A^c \cup B^c)$ (*Theorem*: $(A \cap B)^c = A^c \cup B^c$)

$$= (A \cap A^c) \cup (A \cap B^c)$$

$$= \emptyset \cup (A - B)$$

$$= A - B.$$

RHS $= (A \cup B) - B = (A \cup B) \cap B^c$

$$= (A \cap B^c) \cup (B \cap B^c)$$

$$= (A \cap B^c) \cup \emptyset$$

$$= (A - B) \cup \emptyset$$

$$= A - B.$$

Since LHS = RHS, we have $A - (A \cap B) = (A \cup B) - B$. □

(c) $f : \Re^+ \cup \{0\} \to \Re^+ \cup \{0\}$ is such that $f(x) = x^2$.

(i) *f is one to one*

Proof: We show that $f(x_1) = f(x_2) \Rightarrow x_1 = x_2$, for $x_1, x_2 \in (\Re^+ \cup \{0\})$.

$f(x_1) = f(x_2) \Rightarrow x_1^2 = x_2^2 \Rightarrow x_1^2 - x_2^2 = 0 \Rightarrow (x_1 - x_2)(x_1 + x_2) = 0$.

Now for $x_1, x_2 \geq 0$, $x_1 + x_2 > 0$, unless $x_1 = x_2 = 0$.

So $(x_1 - x_2)(x_1 + x_2) = 0 \Rightarrow x_1 - x_2 = 0 \Rightarrow x_1 = x_2$.

∴ *f* is one to one.

(ii) *Onto*

Proof: We show for every $b \in (\Re^+ \cup \{0\})$, there is an $x \in (\Re^+ \cup \{0\})$, such that $f(x) = b$.

Let $b \in (\Re^+ \cup \{0\})$. $f(x) = b \Leftrightarrow x^2 = b \Leftrightarrow x = \pm\sqrt{b}$.

Now $+\sqrt{b} \in (\Re^+ \cup \{0\})$ for every $b \in (\Re^+ \cup \{0\})$.

So for every $b \in$ codomain, we have an $x \in$ domain, namely $x = +\sqrt{b}$, such that $f(x) = b$.

∴ *f* is onto.

Since f is one to one and onto it is bijective. Since f is bijective it has an inverse: $f^{-1}: \Re^+ \cup \{0\} \to \Re^+ \cup \{0\}$ is such that $f^{-1}(x) = +\sqrt{x}$. □

SECTION B

3. (a) $x * y = x + y + xy, \quad \forall x, y \in \Re$

(i) *Commutative*

$x * y = x + y + xy$

$\quad = y + x + yx \quad$ (+ and · are commutative)

∴ $x * y = y * x, \quad \forall x, y \in \Re$. * is commutative.

(ii). *Associative*

$(x * y) * z = (x + y + xy) * z = x + y + xy + z + (x + y + xy)z$

$\qquad = x + y + xy + z + xz + yz + xyz$

$x * (y * z) = x * (y + z + yz) = x + y + z + yz + x(y + z + yz))$

$\qquad = x + y + z + yz + xy + xz + xyz$

∴ $(x * y) * z = x * (y * z), \quad \forall x, y, z \in \Re$.

∴ * is associative.

(iii) $\Re$ has an identity e iff $e * x = x + e = x, \forall x \in \Re$.

Since * is commutative, we need only check that $e * x = x, \forall x \in \Re$.

$e * x = e + x + ex = x \Leftrightarrow e\ (1 + x) = 0 \Leftrightarrow e = \dfrac{0}{1+x} = 0, \quad x \neq -1$.

But when $x = -1$, $\quad 0 * -1 = 0 + -1 + 0(-1) = -1$.

So, $\forall x \in \Re, \ 0 * x = x$. ∴ 0 is the identity element.

(iv) x^{-1} is the inverse of an element x, if $x * x^{-1} = e = x^{-1} * x, \quad x^{-1} \in \Re$.

Since * is commutative, we need only have $x * x^{-1} = e$.

$x * x^{-1} = x + x^{-1} + x\,x^{-1} = 0 \Leftrightarrow x^{-1}(1 + x) = -x$

$\Leftrightarrow x^{-1} = \dfrac{-x}{1+x}$. If $x = -1$, x^{-1} is undefined.

So every real number x, except $x = -1$, has an inverse under $*$, namely $x^{-1} = \dfrac{-x}{1+x}$.

(b) R is a relation defined on Z, as follows:

$$\forall\, m, n \in Z, \quad m\, R\, n \Leftrightarrow m^2 - n^2 \text{ is even.}$$

R is an equivalence relation.

Proof:

(i) $m\, R\, m$, $\forall\, m$ since $m^2 - m^2 = 0$, which is even. $\therefore$ R is reflexive.

(ii) $m\, R\, n \;\Rightarrow\; m^2 - n^2 = 2t_1$, where t_1 is an integer.

Now $n^2 - m^2 = n^2 - (2t_1 + n^2) = n^2 - 2t_1 - n^2 = -2t_1 = 2t_2$, where $t_2 = -t_1$ is an integer.

$\therefore n^2 - m^2$ is even. Thus $n\, R\, m$. $\therefore$ R is symmetric.

(iii) If $m\, R\, n$ and $n\, R\, p$ then $m^2 - n^2 = 2r$ and $n^2 - p^2 = 2s$.

Now $m^2 - p^2 = m^2 - n^2 + n^2 - p^2 = 2r + 2s = 2(r+s) = 2t$, where $t \in Z$.

Thus $m\, R\, p$. $\therefore$ R is transitive.

Since R is reflexive, symmetric and transitive, it is an equivalence relation. □

The equivalence classes of R:

$[0] = \{x : x\, R\, 0, x \in Z\} = \{x : x^2 - 0 \text{ is even, } x \in Z\}$

$ = \{x : x \text{ is even and } x \in Z\} = \{0, \pm 2, \pm 4, \pm 6, \ldots\}$.

$[1] = \{x : x\, R\, 1, x \in Z\} = \{x : x^2 - 1 \text{ is even, } x \in Z\}$

$ = \{x : x^2 \text{ is odd and } x \in Z\} = \{\pm 1, \pm 3, \pm 5, \ldots\}$.

4. (a) If $x \neq \pm 3$ then either (i) $x > 3$ or (ii) $-3 < x < 3$ or (iii) $x < -3$.

Case (i): If $x > 3$ then $\dfrac{x-3}{|x^2 - 9|} = \dfrac{x-3}{x^2 - 9} = \dfrac{1}{x+3} < 1 \Leftrightarrow x + 3 > 1 \Leftrightarrow x > -2$.

This is always true when $x > 3$. So the given inequality holds for all $x > 3$.

Case (ii): $|x^2 - 9| = 9 - x^2$ and $x > -3$, $\therefore \dfrac{x-3}{9-x^2} = \dfrac{-1}{x+3} < 1 \Leftrightarrow -1 < x+3 \Leftrightarrow -4 < x$.

Hence the given inequality holds for $-3 < x < 3$.

Case (iii): $x < -3 \Leftrightarrow |x^2 - 9| = x^2 - 9$ and $\dfrac{x-3}{x^2-9} = \dfrac{1}{x+3} < 1 \Leftrightarrow x < -2$. This is true for all $x < -3$. So the given inequality holds for all $x < -3$.

Hence $\dfrac{x-3}{|x^2-9|} < 1$ for all $x \neq \pm 3$.

(b) Let $x \geq 0$, $y \in \Re$, then $(\sqrt{x} - y)^2 \geq 0$

$\Rightarrow x - 2\sqrt{x}\, y + y^2 \geq 0$

$\Rightarrow x + y^2 \geq 2\sqrt{x}\, y$. □

(c) We know that $-|a| \leq a \leq |a|$ and $-|b| \leq b \leq |b|$.

Adding these inequalities gives: $-|a| - |b| \leq a + b \leq |a| + |b|$

And so, by the property of absolute value, $|a+b| \leq |a| + |b|$. □

(d) Let $P(n)$ be the statement that $\displaystyle\sum_{i=0}^{n} \alpha^i = \dfrac{1 - \alpha^{n+1}}{1 - \alpha}$, where $\alpha \in \Re$ is given and $\alpha \neq 1$.

Base case: $n = 0$: $\displaystyle\sum_{i=0}^{0} \alpha^i = 1$ and $\dfrac{1 - \alpha^{0+1}}{1 - \alpha} = \dfrac{1 - \alpha}{1 - \alpha} = 1$. Hence $P(0)$ holds.

Induction hypothesis: Assume that $P(k)$ holds, where $k \in N$; that is, $\displaystyle\sum_{i=0}^{k} \alpha^i = \dfrac{1 - \alpha^{k+1}}{1 - \alpha}$.

Induction step: $\displaystyle\sum_{i=0}^{k+1} \alpha^i = \sum_{i=0}^{k} \alpha^i + \alpha^{k+1}$

$= \dfrac{1 - \alpha^{k+1}}{1 - \alpha} + \alpha^{k+1}$ by induction hypothesis

$$= \frac{1-\alpha^{k+1}+\alpha^{k+1}(1-\alpha)}{1-\alpha} = \frac{1-\alpha^{k+2}}{1-\alpha}. \therefore P(k+1) \text{ holds.}$$

By the principle of mathematical induction $\sum_{i=0}^{n} \alpha^i = \frac{1-\alpha^{n+1}}{1-\alpha}$ for all $n \in N$ and $\alpha \neq 1$. $\square$

SECTION C

5. (a) Apply the Euclidean Algorithm:

$510 = 1.310 + 200$

$310 = 1.200 + 110$

$200 = 1.110 + 90$

$110 = 1.90 + 20$

$20 = 2.10 + 0$. Stop.

Hence $(510, 310) = 10$.

Since $10|30$, the equation $310x + 510y = 30$ has infinitely many integer solutions.

We use back substitution to find x, y such that $310x + 510y = 10$:

$10 = 90 - 4 \cdot 20$

$= 90 - 4 (110 - 90) = 5 \cdot 90 - 4 \cdot 110$

$= 5 \cdot (200 - 110) - 4 \cdot 110 = 5 \cdot 200 - 9 \cdot 110$

$= 5 \cdot 200 - 9 (310 - 200) = 14 \cdot 200 - 9 \cdot 310$

$= 14 (510 - 310) - 9 \cdot 310 = 14 \cdot 510 - 23 \cdot 310$

A particular solution is $x = -23, \ y = 14$.

Hence a solution to $310x + 510y = 30$ is given by $x = -23 \cdot 3 = -69, \ y = 14 \cdot 3 = 42$.

The set of all solutions to $310x + 510y = 30$ is $\{(x,y) = (-69 + 51n, \ 42 - 31n) \mid n \in Z\}$.

(b) *Proof* (by contradiction)

Assume that $\sqrt{2} \in Q$; that is, there are integers $x, y \in Z$ with $y \neq 0$ and $\sqrt{2} = \frac{x}{y}$, where $(x, y) = 1$. This implies that $2 = \frac{x^2}{y^2} \Rightarrow y^2 2 = x^2$.

So $2\big|x^2$, which means that x must be even. Write $x = 2n$ for some $n \in Z$.

So $2y^2 = 4n^2 \Rightarrow y^2 = 2n^2 \Rightarrow 2\big|y^2 \Rightarrow 2\big|y$.

So $2\big|y$, $2\big|x$, hence $(x, y) \geq 2$. This contradicts the assumption that $(x, y) = 1$.

Hence x, y do not exist; and so $\sqrt{2} \notin Q$. □

(c) Let $x, y \in Z$ and let $x^2 \big| y^2$

If $y = 0$, then $x \big| y$; since $0 = 0 \cdot x$.

If $y \neq 0$, then as there exists $k \in Z$ with $y^2 = k\,x^2$, we have that $x \neq 0$ as well.

Now factor x, y into a product of primes:

$y = \pm p_1^{a_1}...p_k^{a_k}$ and $x = \pm p_1^{b_1}...p_k^{b_k}$ where $p_1,..., p_k$ are all distinct and all exponents $a_1, a_2,...,a_k, b_1, b_2,...,b_k \geq 0$.

As $x^2 \big| y^2$ and $x^2 = p_1^{2b_1}...p_k^{2b_k}$, and $y^2 = p_1^{2a_1}...p_k^{2a_k}$, we have that

$$\left.\begin{array}{l} 2b_1 \leq 2a_1 \\ 2b_2 \leq 2a_2 \\ \vdots \\ 2b_k \leq 2a_k \end{array}\right\} \Rightarrow \begin{array}{l} b_1 \leq a_1 \\ b_2 \leq a_2 \\ \vdots \\ b_k \leq a_k \end{array}$$

∴ $b_i \leq a_i$ for each $1 \leq i \leq k$. By unique factorization $x \big| y$. □

6. (a). $\dfrac{z-i}{z+1} = \dfrac{x+iy-i}{(x+1)+iy} \cdot \dfrac{(x+1)-iy}{(x+1)-iy}$. (NB: You may leave this question for next semester.)

Multiplying out and setting the imaginary part to be zero gives: $-iyx + iyx + iy - ix - i = 0$.

This gives $y - x - 1 = 0$, i.e. $y = x + 1$. This is the path traced out by z in the Argand diagram.

(b) $\sqrt{3} + i = 2\left(\cos\dfrac{\pi}{6} + i\sin\dfrac{\pi}{6}\right)$, $1 - i = \sqrt{2}\left(\cos\dfrac{\pi}{4} + i\sin\dfrac{\pi}{4}\right)$.

$$\left|\frac{(\sqrt{3}+i)^{17}}{(1-i)^{10}}\right| = \frac{|\sqrt{3}+i|^{17}}{|1-i|^{10}} = \frac{2^{17}}{(\sqrt{2})^{10}} = 2^{12}.$$

Argument $= 17 \cdot \frac{\pi}{6} - \left(10(-\frac{\pi}{4})\right) = 17\frac{\pi}{6} + 5\frac{\pi}{2} = 16\frac{\pi}{3} = \left(\frac{4\pi}{3} + 4\pi\right).$

$\Rightarrow$ argument $= \frac{4\pi}{3}$ and principal value $= -\frac{2\pi}{3}$.

(c) $|z| = 1 \Rightarrow |\bar{z}| = 1$.

$$\left|\frac{z-w}{1-w\bar{z}}\right| = \frac{|z-w|}{|1-w\bar{z}|}|\bar{z}| = \frac{|z\bar{z}-w\bar{z}|}{|1-w\bar{z}|} = \frac{||z|^2 - w\bar{z}|}{|1-w\bar{z}|} = \frac{|1-w\bar{z}|}{|1-w\bar{z}|} = 1. \square$$

(d) Solving $x^3 = 1 \Rightarrow (x-1)(x^2 + x + 1) = 0$.

$\therefore x = 1$ or $x = \frac{-1 \pm \sqrt{1-4}}{2} = \frac{-1 \pm i\sqrt{3}}{2}$.

SOLUTIONS TO DECEMBER 2013–MATH 1152

SECTION A

1. (a) Let A: "I will go," B: "You will go."

The given statements, in the language of logic, are:

i. $A \to B$ ii. $B \to A$ iii. $\sim B \to \sim A$ iv. $A \land \sim B$ v. $B \lor \sim A$.

A	B	$A \to B$	$B \to A$	$\sim B$	$\sim A$	$\sim B \to \sim A$	$A \land \sim B$	$B \lor \sim A$
T	T	T	T	F	F	T	F	T
T	F	F	T	T	F	F	T	F
F	T	T	F	F	T	T	F	T
F	F	T	T	T	T	T	F	T

The truth table shows that statements (i), (iii) and (v) are logically equivalent.

(b) Given statement: *'Some truths are absolute."*

Let $t(x)$ be statement 'truth x is absolute.'

Given statement can be written: $\exists x : t(x)$

Negation: $\sim(\exists x : t(x)) \equiv \forall x, \sim t(x)$

i.e. "There are no absolute truths."

If the given statement is true then its negation is false.

(c)
$$\begin{array}{l} \text{If I speak, I am condemned} \\ \underline{\text{If I stay silent, I am damned}} \\ \text{If I am not condemned, I am damned} \end{array}$$

Let p : I speak, q : I am condemned, r : I am damned.

Given argument can be written: $\begin{array}{l} P_1 : p \to q \\ \underline{P_2 : \sim p \to r} \\ C : \sim q \to r \end{array}$

Since $(p \to q) \equiv \sim q \to \sim p$, the argument can be rewritten $\begin{array}{l} \sim q \to \sim p \\ \underline{\sim p \to r} \\ \sim q \to r \end{array}$

This is classical hypothetical syllogism of the form $\begin{array}{l} A \to B \\ \underline{B \to C} \\ A \to C \end{array}$

The argument is therefore valid. [You may also use truth tables.]

(d) We know that x is not less than y, to prove that $x+1$ is not less than $y+1$:

Proof (by contradiction): Suppose not.

$x+1$ is less than $y+1 \to x$ is less than y

$\underline{\text{But } x \text{ is not less than } y}$

$\therefore$ It cannot be the case that $x+1$ is less than $y+1$

$\therefore x+1$ is not less than $y+1$.

The proof is of the form
$$\begin{array}{c} \sim A \to B \\ \sim B \\ \hline \therefore \sim(\sim A) \end{array}$$
and is a valid proof by contradiction. □

2. (a) Required to prove that $A \subseteq B$ and $B \subseteq C \to A \subseteq C$.

Proof (formal):

Let $x \in A$ where x is arbitrary.

$x \in A \to x \in B$ (since $A \subseteq B$)

and $x \in B \to x \in C$ (since $B \subseteq C$)

$\therefore x \in A \to x \in C$. $\therefore A \subseteq C$.

$\therefore A \subseteq B$ and $B \subseteq C \to A \subseteq C$. □

(b) $(A \cup B)^c \cup (A - B) = B^c$

Proof (set algebra):

$(A \cup B)^c \cup (A - B) = (A^c \cap B^c) \cup (A - B)$, by De Morgan's Law

$= (A^c \cap B^c) \cup (A \cap B^c)$, since $X - Y = X \cap Y^c$

$= (A^c \cup A) \cap B^c$ (intersection distributes over union)

$= U \cap B^c$ (since $(A^c \cup A) = U$).

$= B^c$. □

(c) Let $A = \{$ politicians $\}$, $B = \{$ those who love power $\}$, $C = \{$ those who love money $\}$.

Given argument:

$$\begin{array}{l} P_1 : A \subseteq B \\ P_2 : B \cap C \neq \emptyset \\ \hline C : A \cap C \neq \emptyset \end{array}$$

The 2nd diagram below shows a case when the premises are true but the conclusion is false. The argument is therefore invalid.

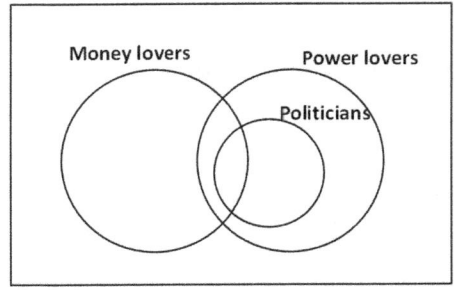
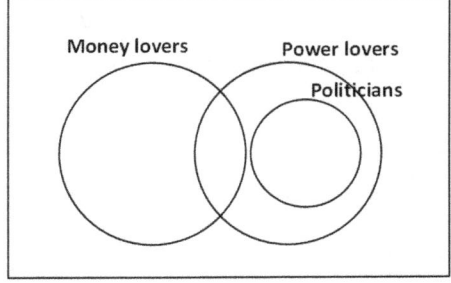

SECTION B

3.(a) $x * y = 1 + \sqrt{(x-1)^2 + (y-1)^2}$, $\quad x, y \geq 1$

(i). *Closure:* $x, y \geq 1 \Rightarrow x - 1 \geq 0, \; y - 1 \geq 0$

$$\Rightarrow (x-1)^2 + (y-1)^2 \geq 0$$

$$\Rightarrow \sqrt{(x-1)^2 + (y-1)^2} \geq 0$$

$$\Rightarrow 1 + \sqrt{(x-1)^2 + (y-1)^2} \geq 1$$

$\therefore x * y \in S$, $\forall x, y \in S$. S is closed under the operation.

(ii). *Commutative:*

$x * y = 1 + \sqrt{(x-1)^2 + (y-1)^2} = 1 + \sqrt{(y-1)^2 + (x-1)^2} = y * x$, $\forall x, y \in S$.

(iii). *Associative:*

$(x * y) * z = (1 + \sqrt{(x-1)^2 + (y-1)^2}) * z$

$= 1 + \sqrt{(x-1)^2 + (y-1)^2 + (z-1)^2}$

$x * (y * z) = x * (1 + \sqrt{(y-1)^2 + (z-1)^2})$

$= 1 + \sqrt{(x-1)^2 + (y-1)^2 + (z-1)^2}$. Therefore $*$ is associative.

(iv). Since $*$ is commutative we need only solve $x * e = x$:

$1 + \sqrt{(x-1)^2 + (e-1)^2} = x$

$\Leftrightarrow \sqrt{(x-1)^2 + (e-1)^2} = x-1$

$\Leftrightarrow (x-1)^2 + (e-1)^2 = (x-1)^2$

$\Leftrightarrow (e-1)^2 = 0$

$\Leftrightarrow e-1 = 0$

$\Leftrightarrow e = 1$. Therefore the identity element is 1.

Inverse: $x * y = 1$

$\Leftrightarrow 1 + \sqrt{(x-1)^2 + (y-1)^2} = 1$

$\Leftrightarrow \sqrt{(x-1)^2 + (y-1)^2} = 0$

$\Leftrightarrow (x-1)^2 + (y-1)^2 = 0$

$\Leftrightarrow (x-1) = 0$ and $(y-1) = 0$ since $a, b \geq 0 \Rightarrow a^2 + b^2 = 0$ *iff* $a = 0 = b$.

$\Leftrightarrow x = 1$ and $y = 1$. $\therefore$ 1 is only element with inverse and $(1)^{-1} = 1$.

(b). $(a,b) \sim (c,d)$ *iff* $a^2 + d^2 = b^2 + c^2$ on $Z \times Z$

(i) ~ is an equivalence relation.

Proof:

$x^2 + y^2 = y^2 + x^2$

$\therefore (x,y) \sim (x,y)$, $\forall (x,y) \in Z \times Z$.

Therefore ~ is *reflexive.*

Now let $(a,b) \sim (c,d)$

i.e. $a^2 + d^2 = b^2 + c^2$

$\therefore c^2 + b^2 = d^2 + a^2$

$\Rightarrow (c,d) \sim (a,b)$. $\therefore$ ~ is *symmetric.*

Suppose that $(a,b) \sim (c,d)$ and $(c,d) \sim (e,f)$

$a^2 + d^2 = b^2 + c^2$ and $c^2 + f^2 = d^2 + e^2$

$\Rightarrow a^2 + d^2 + c^2 + f^2 = b^2 + c^2 + d^2 + e^2$

$a^2 + f^2 = b^2 + e^2$

$\Rightarrow (a,b) \sim (e,f)$. $\therefore \sim$ is *transitive*.

$\therefore \sim$ is an equivalence relation. □

(ii). $[(1,2)] = \{ (x,y) \in Z \times Z : (1,2) \sim (x,y) \}$

$= \{ (x,y) \in Z \times Z : 1^2 + y^2 = 2^2 + x^2 \}$

$= \{ (x,y) \in Z \times Z : y^2 = 3 + x^2 \}$

$= \{ (x,y) \in Z \times Z : y = \sqrt{3 + x^2} \}$

$\therefore [(1,2)] = \{ (-1,2), (1,2), (1,-2), (-1,-2) \}$.

4. (a) $1 \cdot 2 + 2 \cdot 3 + 3 \cdot 4 + \ldots + n(n+1) = \dfrac{n(n+1)(n+2)}{3}$, $\quad n \geq 1$

Proof: Induction on n

Let $n = 1$. L.H.S. $= 1 \cdot 2 = 2$. R.H.S. $= \dfrac{1(2)(3)}{3} = 2$. Theorem is true for $n = 1$.

Assume that $1 \cdot 2 + 2 \cdot 3 + \ldots + k(k+1) = \dfrac{k(k+1)(k+2)}{3}$, for some positive integer k.

Required to show: $1 \cdot 2 + 2 \cdot 3 + \ldots + k(k+1) + (k+1)(k+2) = \dfrac{(k+1)(k+2)(k+3)}{3}$.

LHS $= \dfrac{k(k+1)(k+2)}{3} + \dfrac{3}{3}(k+1)(k+2)$

$= \dfrac{(k+1)(k+2)[k+3]}{3} =$ R.H.S. $\therefore$ Theorem is true for all positive integers. □

(b). $f(x)=\begin{cases} x^2, & x\geq 0 \\ x-1, & x<0 \end{cases}$ $g(x)=(x+1)$

$dom(fog) = dom\, g = \Re$

$(fog)(x)=\begin{cases} (x+1)^2, & x+1\geq 0 \;(\text{i.e. }x\geq -1) \\ (x+1)-1, & x+1<0 \;(\text{i.e. }x<-1) \end{cases}$

$\therefore (fog)(x)=\begin{cases} (x+1)^2, & x\geq -1 \\ x, & x<-1 \end{cases}$

(c). $A\subseteq \Re,\ B\subseteq \Re.\ f: A\to B$ is defined by $f(x)=\dfrac{x-3}{2x+1}$.

(i). No. If $2x=-1$, i.e. $x=-\dfrac{1}{2}$, then f is undefined.

(ii). $A = \Re - \left\{-\dfrac{1}{2}\right\}$ is the largest set for which f is a function.

(iii). *One to One:*

Let $f(x_1)=f(x_2),\ \ x_1, x_2 \in A$.

$\dfrac{x_1-3}{2x_1+1} = \dfrac{x_2-3}{2x_2+1},\ \ x_1, x_2 \neq -\dfrac{1}{2}$

$\Rightarrow (x_1-3)(2x_2+1) = (2x_1+1)(x_2-3)$

$\Rightarrow 2x_1x_2 + x_1 - 6x_2 - 3 = 2x_1x_2 - 6x_1 + x_2 - 3$

$\therefore x_1 + 6x_1 = x_1 + 6x_2$

$\therefore x_1 = x_2 .\therefore f$ is 1-1.

(iv). The largest B for which f is onto is when $B =$ range f.

Now, $f(x) = b \Leftrightarrow \dfrac{x-3}{2x+1} = \dfrac{b}{1}$

$\Leftrightarrow x - 3 = b(2x + 1)$

$\Leftrightarrow x - 2bx = b + 3$

$\Leftrightarrow x(1 - 2b) = b + 3$

$\Leftrightarrow x = \dfrac{b+3}{1-2b}$. This is undefined if $1 - 2b = 0$, i.e. $b = \dfrac{1}{2}$.

$\therefore B = \Re - \left\{\dfrac{1}{2}\right\}$ is the largest set for which f is both 1-1 and onto.

(v). $f^{-1} : \Re - \left\{-\dfrac{1}{2}\right\} \to \Re - \left\{\dfrac{1}{2}\right\}$ is defined by $f^{-1}(x) = \dfrac{x-3}{1-2x}$.

SECTION C

5. (a) $x^2 - 3|x - 1| > 1$

Case 1: $x \geq 1$ or $x - 1 \geq 0$. In this case $|x - 1| = x - 1$.

$x^2 - 3(x - 1) > 1$

$\leftrightarrow x^2 - 3x + 2 > 0$

$\leftrightarrow (x - 2)(x - 1) > 0$

$\therefore x < 1$ or $x > 2$, but also $x \geq 1$.

$\therefore x > 2$.

Case 2: $x < 1$ or $x - 1 < 0$. In this case $|x - 1| = -(x - 1)$.

$x^2 + 3(x - 1) > 1$

$\leftrightarrow x^2 + 3x - 4 > 0$

$\leftrightarrow (x + 4)(x - 1) > 0$

$\therefore x < -4$ or $x > 1$, and also $x < 1$ (for case 2).

$\therefore x < -4$.

Solution set $= \{x : x < -4 \text{ or } x > 2\}$, i.e. $(-\infty, -4) \cup (2, \infty)$.

(b). $\dfrac{x^2}{y^2}+\dfrac{y^2}{x^2}+6\geq \dfrac{4x}{y}+\dfrac{4y}{x}, \quad x,y>0$

Proof

$\dfrac{x^2}{y^2}+\dfrac{y^2}{x^2}+6\geq \dfrac{4x}{y}+\dfrac{4y}{x}$

$\Leftrightarrow \dfrac{x^4+y^4+6x^2y^2}{y^2x^2}\geq \dfrac{4x^3y+4xy^3}{x^2y^2}$

$\Leftrightarrow x^4+y^4+6x^2y^2\geq 4x^3y+4xy^3$

$\Leftrightarrow x^4+y^4+6x^2y^2-4x^3y-4xy^3\geq 0$

$\Leftrightarrow (x-y)^4\geq 0$ (from binomial expansion*).

We know that $(x-y)^4\geq 0$, $\therefore \dfrac{x^2}{y^2}+\dfrac{y^2}{x^2}+6\geq \dfrac{4x}{y}+\dfrac{4y}{x}$. □

(NB $(x-y)^4 = x^4 + \dfrac{4.3.2}{1.2.3}x^3(-y) + \dfrac{4.3.2}{1.2.3}x^2(-y^2) + 4x(-y)^3 + (-y)^4 = x^4 - 4x^3y + 6x^2y^2 - 4xy^3 + y^4$).

(c). $|a-b|\geq |a|-|b|$

Proof:

$|a-b|^2 = (a-b)^2 = a^2+b^2-2ab \geq a^2+b^2-2|ab| = |a|^2+|b|^2-2|a||b| = (|a|-|b|)^2$

$\therefore |a-b|\geq |a|-|b|$ since $|a-b|\geq 0$. □

6. (a) $a,b,c \in Z$, $a|b$ and $b|c \to a|c$

Proof:

Let $a|b$ and $b|c$

$\therefore b=k_1 a$ and $c=k_2 a$ where $k_1, k_2 \in Z$.

$\therefore c=k_2 k_1 a = ka$ where $k=k_1 k_2 \in Z$

$\therefore a|c$. □

(b). $21x + 14y = 147$

Applying the Euclidean Algorithm to find g.c.d. of 21 and 14 we have:

$21 = 1.14 + 7$ (1)

$14 = 2.7$ (2)

Thus $(21, 14) = 7$.

Now $7.21 = 147$, so the Diophantine equation has an infinite number of solutions.

Now by (1), $7 = 1.21 + (-1) \cdot 14$

$$\Rightarrow 147 = (21) \cdot 21 + (-21) \cdot 14.$$

Thus $x_0 = 21$, $y_0 = -21$ is a particular solution of the given Diophantine equation.

Hence, the general form of the solution is:

$$x = 21 + \frac{14}{(21,14)} n_1 \text{ and } y = (-21) - \frac{21}{(21,14)} n_2, \text{ where } n_1, n_2 \in Z.$$

Thus $x = 21 + 2n_1$, $y = -21 - 3n_2$, for $n_1, n_2 \in Z$.

(c). Let n be a positive prime integer. Prove that $\sqrt{n}$ is irrational.

Proof (by contradiction):

Let us assume that $\sqrt{n}$ is rational. Then $\exists$ co-prime integers x and y (gcd = 1), such that $\sqrt{n} = \frac{x}{y}$.

$$\Rightarrow n = \frac{x^2}{y^2} \Rightarrow x^2 = ny^2.$$

Since $n | ny^2$ then $n | x^2$. Since n is prime then $n | x$.

Put $x = nz$ where $z \in Z$. Then $x^2 = n^2 z^2$. So $n^2 z^2 = ny^2 \Rightarrow nz^2 = y^2 \Rightarrow n | y^2$ and therefore $n | y$ since n is prime. Hence $n | x$ and $n | y$. This is a contradiction since x and y are co-prime. Therefore our assumption is false. $\sqrt{n}$ is irrational. $\square$

(d). $\exists\ k,l \in Z$ such that $m = kn$ and $n = lm$.

$\therefore\ m = kn = klm \Rightarrow m - lkm = 0$; i.e. $m(1-kl) = 0$.

Either $m = 0$ or $kl = 1$. If $m = 0$ then $n = 0$, $l = 0$ and $|m| = |n| = 0$.

If $m \neq 0$ then $kl = 1$ and since $k,l \in Z$, we must have either $k = l = 1$ or $k = l = -1$.

In both cases $m = n$ or $m = -n$ which gives $|m| = |n|$ □

SOLUTIONS TO DECEMBER 2014–MATH 1152

---

1. (a) Given Statement: "If it is big then it must be heavy."

   (i) Converse: "If it is heavy then it must be big."

   (ii) Contrapositive: "If it is not heavy then it is not big."

   (iii) Negation: "It is big but it is not heavy."

   (iv) "Being heavy is necessary for being big."

(b)

p	q	$p \Rightarrow q$	$q \Rightarrow (p \Rightarrow q)$	$p \Rightarrow (p \Rightarrow q)$
T	T	T	T	T
T	F	F	T	F
F	T	T	T	T
F	F	T	T	T

The truth values for $q \Rightarrow (p \Rightarrow q)$ and $p \Rightarrow (p \Rightarrow q)$ are different in row 2 of the table.

Hence $q \Rightarrow (p \Rightarrow q)$ is *not* logically equivalent to $p \Rightarrow (p \Rightarrow q)$.

(c) Premises 1 and 2 imply that $q \Rightarrow \sim r$ (hypothetical syllogism). Hence the argument simplifies to:

$q \Rightarrow \sim r$

$\underline{q}$ This is a modus ponens argument. Hence it is valid.

$\sim r$

Alternative solution (Direct Analysis):

Given argument:

$P_1 : q \Rightarrow \sim p$

$P_2 : \sim p \Rightarrow \sim r$

$\underline{P_3 : q}$

$C : \sim r$

Assume that the conclusion is false i.e. $\sim r$ is false.

For P_2 to be true when $\sim r$ is false, $\sim p$ must be false.

For P_1 to be true when $\sim p$ is false, q must be false.

If q is false, then P_3 is false.

Hence there can never be a case in which all the premises are true and the conclusion false. Thus the argument is valid.

(d) R.T. P. by contradiction that if $x^3 + x = 0$, $x \in \Re$, then $x = 0$.

Proof: Assume that $x^3 + x = 0$ and $x \neq 0$.

Now $x^3 + x = 0$

$\Rightarrow x(x^2 + 1) = 0$

$\Rightarrow$ either $x = 0$ or $x^2 + 1 = 0$

Since we assume $x \neq 0$ then $x^2 + 1 = 0$

$\Rightarrow x^2 = -1$

$\Rightarrow x = \sqrt[+]{-1}$

$\Rightarrow x = \pm i$

But $x \in \Re$. This is a contradiction.

Hence our assumption that $x \neq 0$ is false. The result follows. □

2. (a) (i) $(A \cap B)^c \subseteq A^c \cup B^c$

Proof: Let x be arbitrary.

$x \in (A \cap B)^c \Rightarrow x \notin (A \cap B)$

$\qquad \Rightarrow x \notin A$ or $x \notin B$

$\qquad \Rightarrow x \in A^c$ or $x \in B^c$

$\qquad \Rightarrow x \in (A^c \cup B^c)$. $\therefore (A \cap B)^c \subseteq A^c \cup B^c$. $\square$

For $(A \cap B)^c = A^c \cup B^c$, we must have further that $A^c \cup B^c \subseteq (A \cap B)^c$.

(ii) $A \cup \emptyset \subseteq A$

Proof (formal):

$x \in (A \cup \emptyset) \Rightarrow x \in A$ or $x \in \emptyset$

$\qquad \Rightarrow x \in A$ (since $x \notin \emptyset$)

$\therefore A \cup \emptyset \subseteq A$. $\square$

(b) $(A \cup B) \cap (A^c \cup B^c) = (A - B) \cup (B - A)$

Proof (Set Algebra):

L.H.S. $= (A \cup B) \cap (A^c \cup B^c)$

$= [(A \cup B) \cap A^c] \cup [(A \cup B) \cap B^c]$ (Intersection distributes over union on left)

$= [(A \cap A^c) \cup (B \cap A^c)] \cup [(A \cap B^c) \cup (B \cap B^c)]$ (Right distributive)

$= [\emptyset \cup (B \cap A^c)] \cup [(A \cap B^c) \cup \emptyset]$

$= (B \cap A^c) \cup (A \cap B^c)$

$= (B - A) \cup (A - B)$

$= (A - B) \cup (B - A) =$ R.H.S. $\square$

(c) (i) An argument is valid if whenever the premises are true, the conclusion is also true.

P_1 : All men are logical

(ii) P_2 : Socrates is logical

C : Socrates is a man

Let M equal the set of men, L equal the set of logical beings, and s represent Socrates.

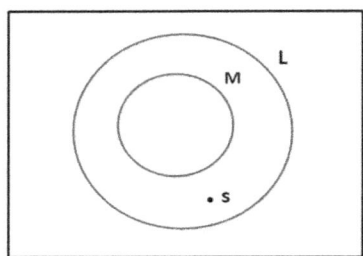

Given argument:

$$P_1 : M \subseteq L$$
$$P_2 : s \in L$$
$$\overline{C : s \in M}$$

The diagram shows a case when the premises are true but the conclusion is not. The argument is therefore invalid.

3. (a) A binary operation $*$ on a set S is a rule which associates with each ordered pair (x, y) in $S \times S$ an element $x * y \in S$.

$$a * b = a + b - 2, \quad a, b \in Z$$

(i) *Commutative*

$b * a = b + a - 2 = a + b - 2 = a * b, \quad \forall a, b \in Z$

$\therefore$ $*$ is commutative

(ii) *Associative*

$(a * b) * c = (a + b - 2) * c = a + b - 2 + c - 2 = a + b + c - 4$

$a * (b * c) = a * (b + c - 2) = a + (b + c - 2) - 2 = a + b + c - 4$

Since $(a * b) * c = a * (b * c), \forall a, b, c \in Z$, $*$ is associative.

(iii) $a * e = a \Leftrightarrow a + e - 2 = a$

$\Leftrightarrow e - 2 = 0$

$\Leftrightarrow e = 2$

Since $*$ is commutative $2 * a = a, \forall a \in Z$ also.

Therefore $e = 2$ is identity element.

(iv) $a * y = 2$

$\Leftrightarrow a + y - 2 = 2$

$\Leftrightarrow a + y = 4$

$\Leftrightarrow y = 4 - a$. Now $(4 - a) \in Z$, $\forall a \in Z$ (since Z is closed under subtraction).

$\therefore$ Every element $a \in Z$ has right inverse, namely, $4 - a$.

Since $*$ is commutative right inverse = left inverse = inverse.

(b) (i) A relation R is reflexive iff $a R a$, $\forall a \in A$.

R is symmetric iff $a R b \rightarrow b R a$

R is transitive iff $a R b$ and $b R c \rightarrow a R c$

(ii) $Z = \{..., -3, -2, -1, 0, 1, 2, ...\}$

$$a R b \text{ iff } a + b \text{ is divisible by 2}$$

Reflexive:

$a + a = 2a$ which is divisible by 2 for all $a \in Z$.

$\therefore a R a$, $\forall a \in Z$. $\therefore R$ is reflexive.

Symmetric:

$a R b \rightarrow a + b = 2z$, for $z \in Z$

$\rightarrow b + a = 2z$

$\rightarrow b R a$. $\therefore R$ is symmetric.

Transitive:

Let $a, b, c \in Z$.

$a R b \rightarrow a + b = 2z_1$, for some $z_1 \in Z$.

$b R c \rightarrow b + c = 2z_2$, for some $z_2 \in Z$.

$\therefore a + b + b + c = 2(z_1 + z_2) = 2z_3$, where $z_3 \in Z$.

$\therefore a+c = 2z_3 - 2b = 2(z_3 - b) = 2z_4$, where $z_4 \in Z$, since Z is closed under $-$.

$\therefore a\,R\,c$. $\therefore R$ is transitive.

Since R is reflexive, symmetric and transitive, it is an equivalence relation. □

$[x] = \{y \in Z : y\,R\,x\}$.

$[0] = \{y \in Z : y + 0 \text{ is divisible by } 2\}$.

$\therefore [0] = \{\ldots, -4, -2, 0, 2, 4, \ldots\}$.

$[1] = \{y \in Z : y + 1 \text{ is divisible by } 2\}$.

$\therefore [1] = \{\ldots, -3, -1, 1, 3, 5, \ldots\}$.

4. (a) Let $P(n)$ be the statement: $\dfrac{1}{1.2} + \dfrac{1}{2.3} + \dfrac{1}{3.4} + \dfrac{1}{4.5} + \ldots + \dfrac{1}{n(n+1)} = \dfrac{n}{n+1}$.

When $n = 1$ we get, $\dfrac{1}{1.2} = \dfrac{1}{1+1}$; i.e. $P(1)$ is true.

Next we assume that $P(k)$ is true, i.e. $\dfrac{1}{1.2} + \dfrac{1}{2.3} + \dfrac{1}{3.4} + \ldots + \dfrac{1}{k(k+1)} = \dfrac{k}{k+1}$.

Now, $\left(\dfrac{1}{1.2} + \dfrac{1}{2.3} + \dfrac{1}{3.4} + \ldots + \dfrac{1}{k(k+1)}\right) + \dfrac{1}{(k+1)(k+2)}$

$= \dfrac{k}{k+1} + \dfrac{1}{(k+1)(k+2)}$

$= \dfrac{k(k+2)+1}{(k+1)(k+2)}$

$= \dfrac{k^2 + 2k + 1}{(k+1)(k+2)} = \dfrac{(k+1)^2}{(k+1)(k+2)}$

$= \dfrac{k+1}{k+2}$. Therefore $P(k+1)$ is true.

The result follows by P.M.I for $n > 0$. □

SOLUTIONS TO PAST EXAM PAPERS

(b) $f : R^+ - \{1\} \to R^+$ is given by $f(x) = \dfrac{x^2}{1+x^2}$

(i) Assume that $f(x) = f(y)$. Then $\dfrac{x^2}{1+x^2} = \dfrac{y^2}{1+y^2}$ $\quad (x, y \neq 1;\ x, y > 0)$

$\Rightarrow x^2(1+y^2) = y^2(1+x^2)$

$\Rightarrow x^2 + x^2 y^2 = y^2 + y^2 x^2$

$\Rightarrow x^2 + x^2 y^2 - y^2 - y^2 x^2 = 0$

$\Rightarrow (x^2 - y^2) = 0$

$\Rightarrow (x - y)(x + y) = 0$

$\Rightarrow x = y$ or $x = -y$

The case $x = -y$ is ignored since x, y are positive. $\therefore x = y$. Thus f is 1-1.

(ii) Let $y \in$ co-domain R^+ (i.e. $y > 0,\ y \neq 0$).

Then $f(x) = y \Leftrightarrow y = \dfrac{x^2}{1+x^2}$

$\Leftrightarrow y(1+x^2) = x^2$

$\Leftrightarrow y + x^2 y = x^2$

$\Leftrightarrow y = x^2 - x^2 y$

$\Leftrightarrow x^2(1-y) = y$

$\Leftrightarrow x = \sqrt{\dfrac{y}{1-y}}$.

x exists provided that $\dfrac{y}{1-y} \geq 0$

$\Leftrightarrow y \geq 0$ and $1 - y > 0$

$\Leftrightarrow 0 < y < 1$ (since $y > 0$). For $y > 1$, there is no x in the domain of f such that $f(x) = y$.

$\therefore f(x)$ is *not* onto.

(iii) $f^{-1}(x) = \sqrt{\dfrac{x}{1-x}}$ for $0 < x < 1$ since $(x \ne 1)$.

5. (a) Required to solve $\dfrac{|x-1|}{1+x} > 0$.

The numerator is always ≥ 0 so we need only consider denominator > 0, i.e. $1 + x > 0$.

$\therefore x > -1$. But if $x = 1$ numerator is zero, therefore the solution is $-1 < x < 1$ or $x > 1$.

(b) Assume $x^3 + y^3 < x^2 y + xy^2$ with $x, y > 0$

$\Rightarrow x^3 + y^3 - x^2 y - xy^2 < 0$

$\Rightarrow x^2(x - y) + y^2(y - x) < 0$

$\Rightarrow (x - y)(x^2 - y^2) < 0$

$\Rightarrow (x - y)(x - y)(x + y) < 0$

$\Rightarrow (x - y)^2 (x + y) < 0$

A contradiction is seen since $(x - y)^2 \ge 0$ and $x + y > 0$.

$\therefore x^3 + y^3 \ge x^2 y + xy^2$.

(c) An integer k is a common divisor of a and b if and only if k divides d, where $d = (a, b)$.

Proof:

Suppose that k is a common divisor for a and b. Since d is the g.c.d. of a and b we can find integers x and y such that $xa + yb = d$.

However $k|a$ and $k|b$, i.e. $a = km$ and $b = kn$ where m, n are integers.

$\therefore xkm + ykn = d$; i.e. $k(xm + yn) = d$, where $(xm + yn) \in Z$.

Hence k divides d.

Conversely if $k|d$ and we have also that $d|a$ and $d|b$, since $d = (a, b)$ then $k|a$ and $k|b$. Therefore k is a common divisor of a and b.

$\therefore$ An integer k is a common divisor of a and b if and only if k divides d. □

SOLUTIONS TO DECEMBER 2015–MATH 1152

1. (a) $(q \Rightarrow p) \Rightarrow (p \vee q)$ is *not* a tautology because its truth values are not all true under the assignment of each variable.

p	q	$p \vee q$	$q \Rightarrow p$	$(q \Rightarrow p) \Rightarrow (p \vee q)$
T	T	T	T	T
T	F	T	T	T
F	T	T	F	T
F	F	F	T	F

(b) Given statement: "Only if Paul goes to school he will get his books and allowance."

This is equivalent to: "Paul will get his books and allowance only if he goes to school."

(i) Given statement as 'if ..., then ...' statement:

"If Paul gets his books and allowance, then he goes to school."

(ii) Let A: 'Paul gets his books and allowance,' B: 'Paul goes to school.'

Statement: $A \to B$, Contrapositive: $\sim B \to \sim A$, converse of contrapositive: $\sim A \to \sim B$.

In words: "If Paul does not get his books and allowance he will not go to school."

But $A \to B$ means A is sufficient for B. So using the word 'sufficient' we have:

"Paul not getting his 'books and allowance' is sufficient for him not going to school."

That is:

"Not getting his books or not getting his allowance is sufficient for Paul not to go to school."

(iii) Given statement: If Bob has two phones then Maria will get a new laptop."

Let A: 'Bob has two phones,' B: 'Maria will get a new laptop.'

Given statement: $A \to B$. Negation: $\sim (A \to B)$. Now $\sim (A \to B) \equiv A \wedge \sim B$.

∴ Negation: "Bob has two phones but Maria will not get a new laptop."

(c). Given argument:

$$P_1 : p \vee s$$
$$P_2 : s \wedge r$$
$$P_3 : p \Rightarrow \sim q$$
$$\underline{P_4 : q \Rightarrow \sim r}$$
$$Con : r \Rightarrow p$$

For the argument to be valid whenever the premises are true the conclusion must be true. Hence the argument is invalid since there is an instance where the premises are true but the conclusion is false:

p	q	r	s	$(p \vee s)$	$(s \wedge r)$	$(p \Rightarrow \sim q)$	$(q \Rightarrow \sim r)$	$(r \Rightarrow p)$
F	F	T	T	T	T	T	T	F

2. (a) (i) RTP by counter example:

$$A \cap (B \cup C) \neq (A \cap B) \cup C$$

Proof: Let $A = \{1, 2, 3, 4\}$, $B = \{1, 2\}$, $C = \{6, 7, 8\}$.

Then, $A \cap B = \{1, 2\}$, $B \cup C = \{1, 2, 6, 7, 8\}$,

$A \cap (B \cup C) = \{1, 2\}$, $(A \cap B) \cup C = \{1, 2, 6, 7, 8\}$.

Therefore, $A \cap (B \cup C) \neq (A \cap B) \cup C$. □

(ii) Diagrams indicate that $A \cap (B \cup C) \subseteq (A \cap B) \cup C$ but $(A \cap B) \cup C \nsubseteq A \cap (B \cup C)$.

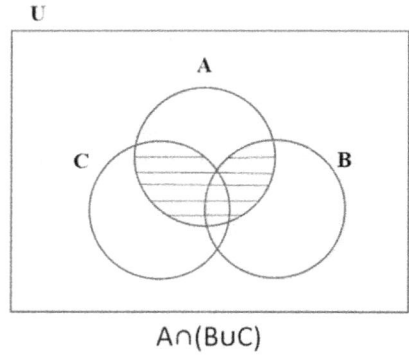

A∩(B∪C)

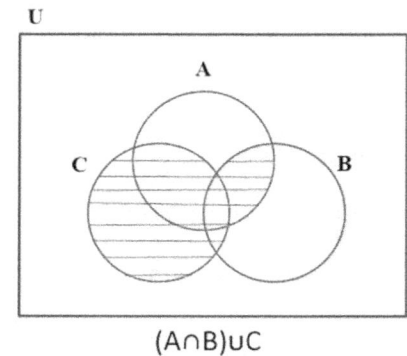

(A∩B)∪C

(A Venn diagram is, however, not a proof! Exercise: Prove $A \cap (B \cup C) \subseteq (A \cap B) \cup C$).

(b) (i) $(A \cap B) \cup \{(A \cap C) \cup (A^c \cap C)\} = (A \cap B) \cup C$

Proof:

$$\text{LHS} = (A \cap B) \cup \{(C \cap A) \cup (C \cap A^c)\} \quad \text{(Intersection is commutative)}$$

$$= (A \cap B) \cup \{C \cap (A \cup A^c)\} \quad \text{(Intersection distributes over union)}$$

$$= (A \cap B) \cup \{C \cap U\} \quad (X \cup X^c = U)$$

$$= (A \cap B) \cup C \quad (X \cap U = X)$$

$\therefore (A \cap B) \cup \{(A \cap C) \cup (A^c \cap C)\} = (A \cap B) \cup C.$ □

(ii) $(A \cap B) \cup \{(A \cap C) \cup (A^c \cap C)\} = \{A \cap (B \cup C)\} \cup (A^c \cap C)$

Proof:

$$\text{LHS} = (A \cap B) \cup \{(A \cap C) \cup (A^c \cap C)\}$$

$$= \{(A \cap B) \cup (A \cap C)\} \cup (A^c \cap C) \quad \text{(Union is associative)}$$

$$= \{A \cap (B \cup C)\} \cup (A^c \cap C) \quad \text{(Intersection distributes over union)}$$

$$= \text{RHS}.$$

$\therefore (A \cap B) \cup \{(A \cap C) \cup (A^c \cap C)\} = \{A \cap (B \cup C)\} \cup (A^c \cap C).$ □

(c) $(A \cap B)^c \subseteq A^c \cup B^c$

Formal Proof:

Let $x \in (A \cap B)^c$

$\Rightarrow x \notin (A \cap B)$

$\Rightarrow x \notin A \text{ or } x \notin B$

$\Rightarrow x \in A^c \text{ or } x \in B^c$

$\Rightarrow x \in (A^c \cup B^c)$

Hence, $(A \cap B)^c \subseteq A^c \cup B^c.$ □

3. (a) $P(X)$ is the set of all subsets of X. $A * B = A \cap B$, where $A, B \in P(X)$.

(i) $A * B = A \cap B$ which is an element of $P(X)$. Thus $P(X)$ is closed under $*$.

(ii) $A * B = A \cap B = B \cap A = B * A$, $\forall A, B \in P(X)$.

$\therefore *$ is commutative.

(iii) $A * (B * C) = A * (B \cap C) = A \cap (B \cap C) = (A \cap B) \cap C = (A * B) * C$, $\forall A, B, C \in P(X)$.

$\therefore *$ is associative.

(iv) If E is identity then $\forall A \in P(X)$, $A * E = A \Leftrightarrow A \cap E = A \Leftrightarrow E = X$. $*$ is commutative,

$\therefore X$ is identity.

(v) Since $*$ is commutative, Y is the inverse of a set A $\Leftrightarrow A * Y = X \Leftrightarrow A \cap Y = X$.

Only $X \cap X = X$. $\therefore X$ is the only element with inverse and the inverse is X itself.

(b) $A = \{a, b, c\}$ and $R = \{(a,b), (b,c), (a,c), (c,c)\}$

(i) $R \subseteq A \times A$, thus R is a relation on A.

(ii) $R^{-1} = \{(b,a), (c,b), (c,a), (c,c)\}$.

(iii) R is not a function since element a is mapped to two images.

(iv) The minimal extension of R to an equivalence relation is:

$$\{(a,b), (b,a), (b,c), (c,b), (a,c), (c,a), (a,a), (b,b), (c,c)\}.$$

The ordered pairs (a,a), (b,b) are included to make the relation reflexive and the pairs $(b,a,)$, (c,b) and (c,a) are needed for symmetry. The relation is also transitive.

(v) Consider $\{(a,a), (b,b), (c,c), (a,b), (b,a), (a,c), (c,a)\}$.

The relation is clearly reflexive and symmetric, however cRa and aRb but (c,b) is not in the relation. Therefore the relation is not transitive.

(vi) Consider $\{(a,a), (b,b), (c,c), (a,b), (b,c)\}$.

Since (a,b) is present but (b,a) is absent the symmetric property does not hold. Also (a,b) and $(b,c,)$ are present but (a,c) is absent so transitivity is spoiled. But each element is related to itself, therefore reflexivity holds.

4. (a) $P_n : \sum_{i=1}^{n}(-1)^{i+1}i^2 = (-1)^{n+1}\dfrac{n(n+1)}{2}$. P_n is true for all positive natural numbers n.

Proof:

$(-1)^{1+1}(1)^2 = (-1)^{1+1}\dfrac{1(1+1)}{2}$. $\therefore P_1$ is true.

Assume P_k is true: $\sum_{i=1}^{k}(-1)^{i+1}i^2 = (-1)^{k+1}\dfrac{k(k+1)}{2}$.

To prove that P_{k+1} is true:

RTP $\sum_{i=1}^{k+1}(-1)^{i+1}i^2 = (-1)^{k+2}\dfrac{(k+1)(k+2)}{2}$.

Proof:

$\text{LHS} = \sum_{i=1}^{k}(-1)^{i+1}i^2 + (k+1)^{th}\text{ term}$

$= (-1)^{k+1}\dfrac{k(k+1)}{2} + (-1)^{k+2}(k+1)^2$

$= \dfrac{(-1)^{k+1}k(k+1) + 2(-1)^{k+2}(k+1)^2}{2}$

$= \dfrac{(-1)^{k+2}\left[(k^2+k)(-1)^{-1} + 2(k+1)^2\right]}{2}$

$= \dfrac{(-1)^{k+2}\left[-k^2 + -k + 2k^2 + 4k + 2\right]}{2}$

$= (-1)^{k+2}\left[\dfrac{k^2 + 3k + 2}{2}\right]$

$= (-1)^{k+2}\dfrac{(k+1)(k+2)}{2} = \text{RHS}.$

$\therefore P_{k+1}$ is true if P_k is true.

Since P_1 is also true, P_n is true $\forall n \in Z^+$. □

(b) $f: R - \{5\} \to R$ is defined by $f(x) = \dfrac{3x}{x-5}$.

(i) Is f one-to-one?

$f(a) = f(b) \Rightarrow \dfrac{3a}{a-5} = \dfrac{3b}{b-5}$

$\Rightarrow 3a(b-5) = 3b(a-5)$

$\Rightarrow 3ab - 15a = 3ba - 15b$

$\Rightarrow -15a = -15b \Rightarrow a = b. \therefore f$ is 1-1.

(ii) Is f onto?

Let $b \in R$. $f(x) = b \Leftrightarrow \dfrac{3x}{x-5} = b \Leftrightarrow 3x = b(x-5)$

$\Leftrightarrow 3x - bx = -5b \Leftrightarrow x(3-b) = -5b$

$\Leftrightarrow x = \dfrac{-5b}{3-b}$. But if $b = 3$, there is no such x in R. Therefore f is *not* onto.

(iii) From (ii), the range of f is $R - \{3\}$.

(iv) Yes, $f^{-1}: R - \{3\} \to R - \{5\}$ exists since f is 1-1.

(v) $f^{-1}(x) = \dfrac{-5x}{3-x}$.

5. (a) $x, y \in \Re^+$

RTP: if $x > y$, then $x^3 + 3xy^2 > 3x^2 y + y^3$.

Consider $(x-y)^3$.

Since $x > y$, $(x-y) > 0$

$\Rightarrow (x-y)^3 > 0$

$\Rightarrow x^3 + 3x^2 y + 3xy^2 - y^3 > 0$

$\Rightarrow x^3 + 3xy^2 > 3x^2 y + y^3$. □

(b) Solve $2x < |x-1|$.

Solution:

Case 1: $x \geq 1$. Then $|x-1| = x-1$.

$2x < |x-1| \Rightarrow 2x < x-1$

$\Rightarrow x < -1$. But $x \geq 1$, so there is no solution in this case.

Case 2: $x < 1$. Then $|x-1| = -(x-1) = 1-x$.

$2x < |x-1| \Rightarrow 2x < 1-x \Rightarrow 3x < 1 \Rightarrow x < \frac{1}{3}$.

So $x < 1$ and $x < \frac{1}{3} \Rightarrow x < \frac{1}{3}$. In this case the solution set $(-\infty, \frac{1}{3})$.

From cases 1 and 2, the solution set is $\{x : x < \frac{1}{3}\}$ —Answer.

(c) $x \equiv y \pmod{m}, \quad x, y \in Z$

$\quad \Rightarrow x - y$ is divisible by m

$\quad \Rightarrow x - y = mk$ for some $k \in Z$

$\quad \Rightarrow ax - ay = amk$, where a is any integer

$\quad \Rightarrow ax - ay = akm$

$\quad \Rightarrow ax - ay = pm$, where $p = ak \in Z$

$\quad \Rightarrow ax - ay$ is divisible by m.

$\therefore ax \equiv ay \pmod{m}$ for any integer a. $\square$

FUTURE MATHEMATICIAN

MATH STUDENT

Without Mathematics the world would not be the same as it is today. She is truly the 'Queen of the Sciences.' Studying Mathematics helps you to develop logical thinking and analytical skills that will serve you in any discipline. A strong background in Mathematics is highly valued by employers and provides opportunity for many fantastic careers. Many students have stifled their love for Mathematics to go into more 'promising' fields, but there is a demand for mathematicians and statisticians across a range of sectors and Mathematics graduates have excellent job opportunities. In fact, CareerCast, an internet job search site, ranked 'Mathematician' as the best job for 2014. It was followed by 'University Professor', 'Statistician' and 'Actuary', in that order. Professions in Mathematics also topped the list for 2015 and 2016.[1]

With a degree in Mathematics you can go into careers in actuarial science, insurance, banking, accounting and other finance related fields. You may further specialise in a particular branch of Mathematics, such as Operations Research, which uses mathematical methods to arrive at optimal solutions to decision problems. For those majoring in Statistics, there is a diverse array of job opportunities in statistical research, including opportunities for collecting and analyzing data for various governmental and private agencies. A joint major with Economics opens careers in econometrics. You might also pursue studies in Bio-Mathematics, Biostatistics, Demographics, Epidemiology and other interesting fields. Mathematics skills are needed in Computer Science, Cryptography, Physics, Engineering and other Math–related fields.

And then there's the most noble of professions—nurturing young minds and helping them to develop their mental powers and problem solving skills, and to love and appreciate Mathematics as much as you do. Mathematics teachers are in high demand! You may go on to earn a Master's or Ph.D. in Pure or Applied Mathematics and go into academia or consulting. You may even, someday, become a Mathematics professor and have a theorem named after you.

If you love Mathematics, pursue it with a passion. It will bring its own reward.

---

[1]. CareerCast, *Jobs rated 2014*, http://www.careercast.com/jobs-rated/jobs-rated-2014-ranking-200-jobs-best-worst. You can learn more about careers in mathematical sciences at http://www.maa.org/careers/career-profiles/sales-marketing/we-do-math. The Mathematical Association of America has also published a book entitled, *101 careers in Mathematics*. You may purchase it on Amazon.

ABOUT THE AUTHOR

Angela Shirley is a lecturer in the Department of Mathematics & Statistics, at The University of West Indies, St Augustine, Trinidad and Tobago. In 1987, she obtained her Ph.D. from Northeastern University, Boston, Massachusetts, with a specialization in Mathematical Statistics. She has been teaching Mathematics ever since—first at Mt. St. Mary's College, Los Angeles, and then at UWI, Mona, before coming to St Augustine in 1996. She has taught courses in Introductory Mathematics; as well as Statistics and a range of other Mathematics courses, both at the undergraduate and graduate level. She also spent three years as a lecturer in Mathematics Education at the School of Education, UWI, St Augustine, and is certified in University Teaching and Learning. Dr. Shirley has a love for imparting mathematics skills to her students and for seeing them succeed. She lives in St. Joseph, Trinidad, with her husband, Lindsay; and has three children.

www.ingramcontent.com/pod-product-compliance
Lightning Source LLC
Chambersburg PA
CBHW080617190526
45169CB00009B/3211